THE MOON'S GALACTIC HISTORY

A Look at the Moon's Extraterrestrial Past and Its Connection to Earth

Constance Victoria Briggs

Adventures Unlimited Press

Acknowledgements

I would like to acknowledge Marcus,

Kion, and Shireen,
Lola Walker, Terry Lamb,
David Hatcher Childress and Adventures Unlimited Press

THE
MOON'S
GALACTIC
HISTORY

Constance Victoria Briggs

The Moon's Galactic History

by Constance Victoria Briggs

ISBN 978-1-948803-50-2

Published by:
Adventures Unlimited Press
One Adventure Place
Kempton, Illinois 60946 USA
auphq@frontiernet.net

www.AdventuresUnlimitedPress.com

THE MOON'S GALACTIC HISTORY

A Look at the Moon's Extraterrestrial Past and Its Connection to Earth

Adventures Unlimited Press

Dedication

To Ghobad
Thank you for encouraging me to write this book!

Other Books by
Constance Victoria Briggs:

The Encyclopedia of Moon Mysteries
Encyclopedia of the Unseen World
The Encyclopedia of Angels
The Encyclopedia of God

TABLE OF CONTENTS

Apollo 16 photo of the far side eartern limb of Moon.

Introduction

Who will tell the Moon's story? That was my thought as I began to write this book. I am the author of several mystical, metaphysical, and spiritual books. However, my studies in cosmic and galactic mysteries, coupled with my writing about the Moon have become my passion. It is as if I have been given a *shot* in life to wake up and write about galactic events and our place in the universe. So, today, it is all about the Moon.

This book takes us on a journey. It is one of mystery, intrigue, and secrets. Throughout history the Moon has mesmerized and mystified us. To some, it represents love and romance. Others view it as an object of scientific wonder, while still others consider it simply a dead rock in the sky. What many do not realize is that the Moon has a fascinating story to be told; one that spans billions of years. It is a tale filled with mystery, intrigue, and secrets. Today, there are researchers, scientists and others that openly admit that there is more to the Moon than meets the eye. There is more to it than what we have been taught. In this book, we follow the past and present research about the Moon's galactic history, with a focus on its connection to Earth and a possible extraterrestrial presence. It examines such questions as, "What is the Moon's origin?" "Are there Moon inhabitants?" "Are the UFOs seen around Earth coming from the Moon?" and more. In the end, we discuss what this all means for *us* and our place in the universe. With all the information on unidentified flying objects in the news today, it is time to have an updated book, with current research, that investigates what is going on with the Moon, and what it means for Earth.

The first chapter, titled *The Moon's Mysterious Ascension,* takes us back to the Moon's history with ancient Earth. It relays accounts of a time when the Moon was not in the sky and ends with an amazing ancient tale of the Moon moving into Earth's orbit. In *Hollow Moon = Spaceship Moon with ET Travelers?* we consider the theory that the Moon is a hollowed-out spaceship. Although this idea has been discussed in the past, this book takes it to another level giving hypothetical scenarios of who or what is occupying the hollow spaceship Moon, and the reasons it may have been brought here. *An Inhabited Alien World* takes us to the early telescopic years and the astronomers' views and theories of that time about the Moon. We then move into hypothetical situations as to what sort of habitation may be there today, relaying several scenarios from there being a Moon colony to the Moon being an actual extraterrestrial country. *Ancient Aliens and the Moon Connection* examines whether the alleged ancient astronauts that visited Earth in the past, may have come from the Moon. We look at tales of space travel during that period and if they were the beings posing at gods during that time.

In *Strange Happenings* we delve into anomalies, and mysterious events on and around the Moon, leading to the question, "Is this evidence of an intelligent race on the Moon?" *A Moon Metropolis!* takes a close look at the possibility of there being a city (or cities) on the Moon. *Hark! Who Goes There?!* gives an account of what extraterrestrial group may be on the Moon. There is a ton of fascinating information today on alleged various extraterrestrial races. Could one of these races be occupying the Moon? *Symbols, Codes, Clues!* If there is an extraterrestrial race on the Moon, are they trying to communicate with us? Have clues been left for us by extraterrestrials in the past? If so, what were they trying to say? Here we will look at possible symbols, codes, clues, and signals left for us from possible extraterrestrials on the Moon, with some possibly here on Earth. In *Where No Man Had Gone Before,* we will probe the strange experiences of the Apollo astronauts during their missions to the Moon. *Into the Future* looks at the future of Moon exploration, space travel, first contact and our place in the universe.

The Earth-Moon system, as seen by the Mars Reconnaissance Orbiter.

Lunar mascon map showing mascons in the five of the largest "seas" on the Moon: Mare Imbrium, Mare Serenitatus, Mare Crisium, Mare Humorum and Mare Nectaris.

The far side of the Moon.

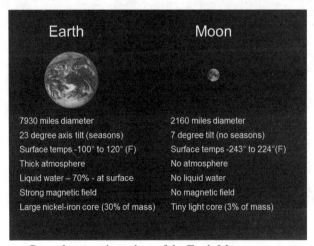

Earth	Moon
7930 miles diameter	2160 miles diameter
23 degree axis tilt (seasons)	7 degree tilt (no seasons)
Surface temps -100° to 120° (F)	Surface temps -243° to 224°(F)
Thick atmosphere	No atmosphere
Liquid water – 70% - at surface	No liquid water
Strong magnetic field	No magnetic field
Large nickel-iron core (30% of mass)	Tiny light core (3% of mass)

General comparison chart of the Earth-Moon system.

Chapter One

The Moon's Mysterious Ascension

"The beginning of wisdom is, I do not know.
I do not know what that is."
—Data *(Star Trek: The Next Generation)*

Our Moon is an anomaly. There is no other moon like it in our Solar System. No other moon behaves the way that it does. Scientists do not have a definitive answer as to how the Moon came to be. There are many theories surrounding its origin. Some are scientific, and others are fantastical. Some believe that it is a dead planetoid harboring life, others believe that it is a spaceship created and brought here by extraterrestrials. Others maintain that the Earth and Moon are a binary planetary system. We simply do not know what the Moon is, how it was created, nor where it came from. All we know for sure is that it does not fit the basic criteria of a natural satellite. So then, if it is not a natural satellite, what is it and where did it come from? If it is a natural satellite, then why is it so difficult for scientists to give a concrete answer about its origin? The popular science fiction writer Isaac Asimov once asked when referring to the Moon, "What is it?" Today, the question of the Moon's origin remains open. There is no definitive answer. The question, "What is it?" persists.

To answer this question, we need to discard what we have been taught about the Moon. I am a firm believer in researching a subject, questioning, and searching for the answers. If I have learned anything from my sojourn into the mysterious subjects of the cosmos, it is that we cannot assume the answers given are all there is. The idea that the Moon is a lifeless dead rock in the sky does not match the mysterious events surrounding it. Something

13

is going on. If mankind is to ever grow, then we must be vigilant in opening our minds to certain truths and abandoning old ideas and teachings. We need to move forward in our examination of our world and the universe. What if we are not alone in the universe? What if extraterrestrials are among us and watching us. What if they were on the Moon? Could all the talk these days about UAP (unidentified aerial phenomena) sightings be connected to the Moon? And if the answer is yes, then how does it affect us? How does it connect to Earth? Probing the possible answers to these questions can be quite exciting. The answers could be life altering, and possibly shift our view of reality and our place in the universe.

The Moon's Origin, *Still* a Mystery

Here we look at the most popular hypotheses proposed by scientists and researchers as to how our Moon originated. We begin with the "co-accretion theory." Sometimes referred to as the "common birth theory," the co-accretion theory is a hypothesis of the Moon's origin that was put forward by the renowned French scholar and astronomer, Pierre-Simon marquis de Laplace (1749-1827). It suggests that the Earth and the Moon formed during the same period, but autonomously of one another, from the same nebulous cloud of dust and gas, which then amalgamated over time. Since the two would have formed from the same nebulous cloud at the same time, that would mean that each should have equal amounts of iron. This is not the case; therefore, this theory was rendered invalid by scientists. The "fission theory" was introduced by George Darwin (the son of English naturalist Charles Darwin). Darwin's hypothesis suggested that the Moon was at one time a part of the Earth. He surmised that, in the beginning, during the forming of the Solar System, the hot, molten Earth spun so rapidly that it became warped. During this course, a piece of the Earth was ripped off and then flung into space, eventually forming the Moon. The Pacific Ocean basin is thought to be the area from which that piece of the Earth was torn. At first, this hypothesis was considered viable since the Moon's makeup is similar to the Earth's mantle. It was believed that a fast-turning Earth could have ejected material to form the Moon from its external strata. However, scientists eventually dismissed this explanation maintaining that this theory

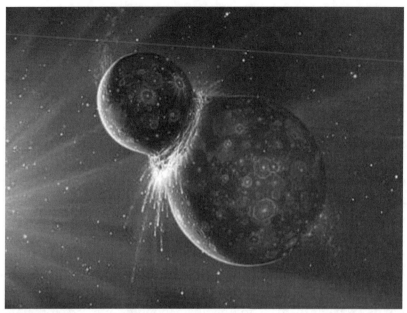

Artist's depiction of the "giant impact" theory of the creation of the Moon. (NASA)

did not provide an acceptable reason for the additional baking that the Moon's material received. One of the more popular theories was the "capture theory." It proposed that the Moon became caught in Earth's gravity after being thrown off its previous course, and thus became Earth's satellite. However, scientists maintained that the Moon is far too large to have become trapped in Earth's orbit, and the theory was eventually dismissed. The "giant impact theory," sometimes referred to as the "big whack theory," is the most widely accepted possibility as to how the Moon was formed. It proposes that the Moon was created when a planet approximately the size of Mars crashed into Earth, triggering large portions of land to tear off, spewing into space, and ultimately creating the Moon. Even though this is the most accepted theory, it is not definitive. The National Aeronautics and Space Administration's (NASA) website, *Curation/Lunar* states, "Recent computer models indicate that the Moon 'could have' been formed from the debris resulting from the Earth being struck a glancing blow by a planetary body about the size of Mars."

Two lesser-known theories are the "asteroid theory" and the "binary planet system theory." The "asteroid theory" proposes that the Moon might be an asteroid that was caught in Earth's gravity, resulting in it becoming a satellite. Another supposition put forward by researchers question whether the Earth and the Moon are really a "binary planet system." This means that they are two planets orbiting each other, whose sizes are so closely matched to one another, that it would be incongruous to refer to them as a planet and a moon. This recalls a quote from American science writer Isaac Asimov where he writes, "What in blazes is our moon doing way out there? It's too far out to be a true satellite of Earth, it's too big to have been captured by the earth. The chances of such a capture having been affected and the moon then having taken up a nearly circular orbit about the earth are too small to make such an eventuality credible. But then, if the moon is neither a true satellite of the earth nor a captured one, what is it?"

With all the theories put forward about the origin of the Moon, coupled with the fact that we have sent several scientific missions to the lunar surface, it is interesting to consider that we still do not have the answers as to how the Moon was created. It remains one of our largest unsolved cosmic mysteries. Perhaps planetary

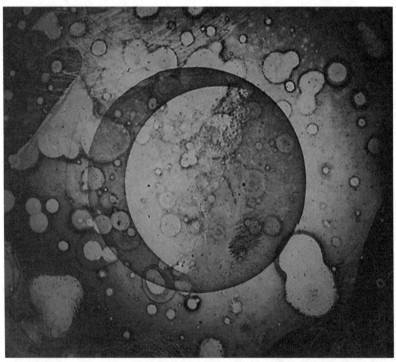

The world's oldest photograph of the Moon, by John William Draper, 1840.

16

scientist William Hartmann said it best when he stated, "Neither the Apollo astronauts, the Luna vehicles, nor all the king's horses and all the king's men could assemble enough data to explain the circumstances of the moon's birth."

Starting Point...

Most people will agree that without the Moon, the Earth would be vastly different. However, despite the amount of scientific information that scientists have collected about the Moon, be it from astronauts that walked on the lunar surface, unmanned space probes, astronomers, and others, there are still vital questions to be answered. Therefore, after all the years of research, time and money spent, the Moon is still a mystery. So much so that it has been labeled "the Rosetta Stone of planets" by American astronomer and planetary physicist Dr. Robert Jastrow. Former *New York Times* science writer Earl Ubell expressed the same line of thought in his article *The Moon Is More of a Mystery than Ever* (April 16, 1972), when he wrote, "The Moon is more complicated than anyone expected; it is not simply a kind of billiard ball frozen in space and time, as many scientists had believed. Few of the fundamental questions have been answered, but the Apollo rocks and recordings have spawned a score of mysteries, a few truly breath-stopping." What many do not realize, is that with the Moon comes a story to be told. It is a tale that spans billions of years; one that quite possibly begins outside of our universe. It is time that we find answers to this mysterious entity. It is time that we understand that the Moon is not just a lifeless rock in the sky but is quite possibly an inhabited alien world!

Tales of a Pre-Lunar Earth

Let us return to the beginning, to a period in Earth's history when it is said that there was no Moon in the sky. According to ancient writings, oral tales and myths from various civilizations, there was a time when only Venus shined in the night sky. It was an age of a pre-lunar Earth. A time before man had the phases of the Moon to tell the seasons, days, and years. It was a period on Earth unlike you and I would ever have imagined and cannot begin to fathom. According to ancient writings of several Greek philosophers, there was a group of people that lived during this period. Just how many tribes of people existed during this era is of course unknown. However, we know of one group that was spoken of by the ancients, that lived in those alleged pre-lunar days. They were called the "Arcadians." Later, as time progressed, they were referred to as the "Proselenes" and the "Pelasgians," which means, "those that lived before the Moon." This was a primitive time when humans were said to have been barbaric, uncultured, and wild. The Arcadians are believed to have existed before the ancient kingdoms of Mesopotamia and Egypt, which existed at the beginning of mankind's history. Mesopotamia and Egypt were there at the cradle of civilization, which would make the Arcadians extremely ancient, going even farther back than mankind's records allow us to remember. In other words, this was a time that we do not find in our history books, except for one very interesting point: this era was mentioned in several writings by well-respected philosophers.

In the modern age, we have been taught that without the Moon, the Earth would be doomed. It is thought that minus the Moon, the Earth would fall into chaos due to it tilting and moving because of a lack of stability on its axis. As a result, it is believed that most of life on Earth would come to a near point of extinction. There are those however, that dispute this, arguing that Mars does not have a Moon even close to the size of Earth's, yet it is not wavering.

There are writings from prominent, ancient Greek philosophers and storytellers whose works have been preserved and handed down to us today on this matter. They should not be ignored. Perhaps their commentaries can help shed light on the notion of a pre-lunar Earth. These men of old wrote of a time when they claimed that the Moon did not exist, a time when there was no moon in the sky. All that was there to illuminate the night sky during that period was the planet Venus. The men that wrote about this era, were the pillars of their society, and their achievements are hailed by us even today. It is difficult to doubt their words since they were renowned for their knowledge, intelligence, and wisdom, and are responsible for helping to educate and mold mankind's understanding of our world and the cosmos. Thus, their words should be taken in a manner of seriousness, no matter how

The Moon ib all its glory.

farfetched people might think the idea of there being no moon in the sky may be. How else are we to completely understand our historic, cosmic past, if we do not revisit the teachings and the writings of these great men that lived at the beginning of mankind's civilization, all the while recording what they saw and experienced during their time on Earth. Therefore, if we accept their works and writings in other subjects, then we cannot discount what they have stated about there being a time in our history when there was no moon. Of course, there are some skeptics that believe that the phrases used by the Greek philosophers when making comments such as, "time without a Moon," are simply colloquialisms. However, we should consider if all the people that spoke of this period were using colloquialisms or not. That is doubtful!

Some of the great thinkers that mentioned such a time in our history include Anaxagoras of Clazomenae, Aristotle, Giordano Bruno, Censorinus, Democritus, Dionysius Chalcidensis,

Hippolytus of Rome, Lucian of Samosata, Mnaseas of Patrae, Plutarch, Ovid, Stephanus of Byzantium, and Theodorus of Cyrene. Publius Ovidius Naso (43 BC–17 AD), more commonly known as Ovid, was a prolific and celebrated Roman poet that lived during the reign of Augustus. His work greatly inspired several illustrious writers in antiquity including Dante, Chaucer, Goethe, Milton, as well as Shakespeare. He is best known for his works *Ars amatoria* and *Metamorphoses*. Ovid mentions a time when the Earth was devoid of a Moon.

He elaborated, giving us a glimpse into the lives of the Arcadians in his work titled *Fasti*. There he writes, "The Arcadians are said to have possessed their land before the birth of Jove [the god Jupiter], and their race is older than the Moon. They lived like beasts, lives spent to no purpose: The common people were crude as yet, without arts. They built houses from leafy branches, grass their crops. Water, scooped in their palms, was nectar to them. No bull panted yoked to the curved ploughshare,. No soil was under the command of the farmer. Horses were not used, all carried their own burdens. The sheep went about still clothed in their wool. People lived in the open and went about nude, inured to heaven downpours from rain-filled winds. To this day the naked priests recall the memory of old customs and testify to those ancient ways." (*Fasti, Book II*).

Ovid's description is so detailed, so precise, that we can easily imagine that time without a moon, where humans were uncultured and feral. The Greek philosopher Democritus (460–370 BCE) was known for developing the concept of the atom and was one of the most outstanding and inexhaustible authors of the ancient world. He too spoke about the Arcadians in one of his works, naming them as a people that existed prior to there being a moon in the sky.

The prominent satirist Lucian of Samosata (120 CE–180 CE), another prolific writer, also mentioned the Arcadians in his work about astrology, where he stated, "The Arcadians affirm in their folly that they are older than the moon." Mnaseas of Patrae, a sought-after scholar and Greek historiographer during the late 3rd century BCE, also wrote about a period when there was no moon. He wrote that there were natives of the land that were referred to as the "Arcadians," and that since they lived prior to the Moon, they were called "Proselenes (people that lived before the Moon)." Plutarch, of Chaeronea in Boeotia (ca 45–120 CE), was a renowned Greek Platonist theorist, and was known to be a prolific writer. He was recognized for his works *Parallel Lives* (about paired Greek and Roman statesmen, as well as military leaders); and

At one time, Venus may have been the brightest object in the night sky.

Moralia, a compilation of moral essays. He also authored *On the Face in the Orb of the Moon.* Plutarch regarded the Moon's topography as being similar to that of the Earth, even though the Moon is smaller. In his essay *The Roman Questions,* he commented on a civilization that existed before the Moon, writing, "There were Arcadians of Evander's following, the so-called pre-Lunar people." Apollonius of Rhodes (295 BC–Unknown) was a major Greek writer of antiquity who was recognized for his work titled *Argonautica.* Apollonius too wrote of an era in Earth's past when the Moon did not exist, even noting that it was during the period preceding the "Danaans" and "Deucalion" civilizations. In *Argonautica IV. 264,* he writes, "When not all the orbs were yet in the heavens, before the Danaans and Deucalion races came into existence, and only the Arcadians lived, of whom it is said that they dwelt on mountains and fed on acorns, before there was a moon." Italian philosopher and cosmologist theorist Giordano Bruno (1548–1600), in his work titled *De Immenso: (Bk IV, x,* pp. 56-57*),* wrote of a pre-lunar period on Earth. Bruno comments, "There are those who have believed that there was a certain time (as our Mythologian says) when the moon, which was believed to be younger than the Sun, was not yet created. The Arcadians, who dwelt not far from the Po, are believed to have been in existence before it [the moon]." Finally, we come to Hippolytus of Rome

The Moon crossing the Earth. (NASA)

(160 AD–236 AD). Hippolytus was an eminent Christian theologist and ecclesiastical author during the second-third century. In his writing titled *Refutatio Omnium Haeresium V. ii.*, Hippolytus wrote one simple line relating to a time without a Moon. He states, "Arcadia brought forth Pelasgus [the first man], he was older than the Moon."

It was not only ancient philosophers from Greece and Rome that spoke of a pre-lunar era. Biblical authors similarly wrote of a period when there was no moon. The Old Testament book of *Job* 25:5 states, "the grandeur of the Lord who 'Makes peace in the heights' is praised and the time is mentioned before [there was] a moon and it did not shine." Additionally, in the Old Testament book of *Psalms* 72:5 it states, "Thou wast feared since [the time of] the sun and before [the time of] the moon, a generation of generations." The Mayans have an ancestral tale that recounts an age when there was no moon, when only the planet Venus lit up the night sky. The native populace of Bogota, Columbia recall their ancient past in the tradition of oral storytelling. They have preserved an ancient legend from a period in their history when the Moon did not exist. At the beginning of the story, it states simply, "In the earliest times, when the moon was not yet in the heavens…" In Tibetan lore, an old tale has been kept and handed down, that speaks of a lost continent named Gondwana. According to the story, Gondwana was home to a culture of people that dwelled there before the Moon appeared.

A Terrible Paradox

All these references concerning an epoch of time when there was no Moon are vital to our understanding exactly what the Moon is, and where it came from. It is interesting to note that while we are contemplating how the Moon originated, we are also opening to an idea that most have never considered before. It forces us to think outside of the box, because suddenly we must stop and consider the unthinkable, that is, the possibility that the Moon was not always there; and if it were not, then where did it come from? The *terrible paradox* is that if we do not take the words of the great philosophers and writers of our past seriously in this matter—those that have taught us so much—if we do not believe the tales preserved and handed down to us, and the scriptures so many claimed to be truth, then perhaps we cannot consider other teachings that we have come to rely on and revere, and live by as well. We cannot simply pick and choose the words of those that have had such great influence on mankind. We cannot turn a blind eye to what is in front of us. There is no in-between. There are no gray areas. We should consider that we have reached a time in mankind's history where we must reevaluate the past. If we do, it appears that we will have to do so painstakingly, piece by piece. One day soon, we just may be rewriting mankind's history; and it may begin with the Moon.

A Tale of the Moon's Arrival

In the ancient city of Tiahuanaco in Bolivia, there is an enormous ruin of a megalithic stone structure. It is known as the Gateway of the Sun (or Gate of the Sun). It is approximately 10 feet tall and is carved from a block of stone that weighs 10 tons. The Gateway of the Sun has strange, mystifying symbols and characters that have been sculpted into it. The Sun-God, Viracocha (the god of creation), is positioned in the center with sunrays radiating from his head. The other figures appear to be a mix between humans and animals. Some have wings, while others have curled tails. Still other figures appear to be wearing head coverings that resemble helmets. Archeologists believe that the Gateway of the Sun was previously used as a calendar; and seems to suggest a solar year.

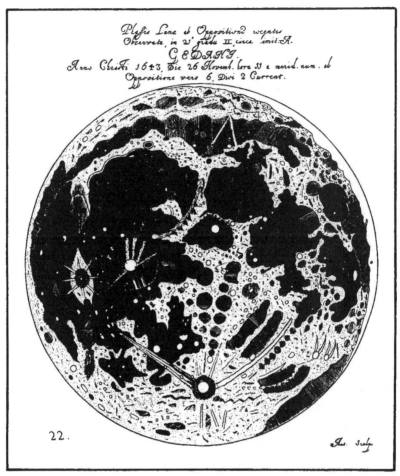

An early map of the Moon made in 1643.

Interestingly, the solar year does not correspond to the solar year that we know today.

According to the symbols, the Moon appeared in our Solar System approximately 12,000 years ago. In the process of its arrival, it created pandemonium on the Earth. This conclusion was reached in 1956 by researcher Hans Schindler Bellamy who had studied the Gateway of the Sun and interpreted the ancient symbols. He wrote about his findings in his work titled *The Calendar of Tiahuanaco*. They reveal an amazing story about the Moon's history. Bellamy explains that the symbols and characters contain mathematical and astronomical knowledge. In addition to revealing when the Moon first arrived, they state that before the

Moon came in, the Earth revolved slower and had only 290 days. Because of this difference in time, it is believed that the Earth was turning on its axis at a slower rate than it is today. Consequently, it had longer days. Bellamy's interpretation of the symbols tells us that these characteristics of Earth changed with the Moon moving in. The Moon's arrival would have caused chaos on Earth because of the gravitational pull. There would have been an upsurge in storms, earthquakes and floods and there would have been a great number of disasters. It is thought that this event may have been the source of the great flood in the tale of Noah's ark, as well as other flood stories of various civilizations around the world. Most interestingly, the Gateway of the Sun had been placed on a platform temple known as Kalasasaya. Kalasasaya is located next to a second temple that is partially underground. Together, the two temples construct part of what is believed to be an ancient observatory. If we are to believe what the symbols reveal, then these ancient people of Bolivia possessed an understanding of the cosmos that we cannot comprehend. Where did they obtain their information?

There have been other tales about older civilizations that had knowledge of the universe that only came to light in our present day. There is an account of a tribe of people out of Africa, named the "Dogon," that had been visited by extraterrestrials and

FRONT OF GREAT MONOLITHIC GATE-WAY.

A tale of the Moon's arrival is said to be found on this Gate of the Sun (in Bolivia). Illustration is from 1877.

givenaastronomical information. The Dogon, as the story goes, knew for centuries that there was a Sirius A and a Sirius B. This was only later confirmed in our modern era. Could it be that one day soon, we will get a confirmation that there was a time without a Moon from the scientific community? The description of the Moon's arrival on the Gateway of the Sun lines up with the ancient writers' comments about a time when there was no Moon in the sky.

For many, the idea of there being a pre-lunar Earth, and ancient symbols telling a story of the Moon moving into Earth's orbit, is sheer fantasy. However, this information opens the conversation that is undoubtedly coming; that is whether we are alone in the universe. We should consider that mankind is a young race compared to the age of the universe, and that there could be beings that may have existed for eons before we appeared. If so, then who knows what sort of advanced technologies and science they may have achieved. Their capabilities would be vastly ahead of ours. Therefore, when we hear tales such as that from the ancient Gateway of the Sun regarding the Moon, we should consider from whence this information came. Could it have come from beings so far advanced than mankind that they visited Earth and brought knowledge of the cosmos to selected ancient cultures? If so, then the Moon has a past that is far beyond what we have come to understand.

In this case, the truth may really be *stranger than fiction...*

Chapter Two

Hollow Moon = Spaceship with ET Travelers?

"For the World is Hollow…and I have touched the sky!"
—*Star Trek* Season 3, Episode 8

Most people have heard the phrase, "the Moon rang like a bell." This became a well-known expression due to an event that occurred during NASA's *Apollo 12* mission. In November 1969, during the *Apollo 12* mission to the Moon, the astronauts set up seismometers on the lunar surface. After they returned to the Command Module, they purposely crashed the Lunar Module's Ascent Stage into the Moon. This occurred approximately forty miles from the *Apollo 12* landing site. The impact with the Moon was of a strength comparable to one ton of TNT. There, the seismometers recorded the reverberations, which astonished NASA officials. The shock waves lasted for approximately one hour. A NASA scientist was quoted as saying that the Moon "rang like a bell."

The Moon resonating for so long could not be explained. Maurice Ewing, one of the directors of the seismic testing, commented during a news conference, "As for the meaning of it, I'd rather not make an interpretation right now, but it is as though someone had struck a bell, say in the belfry of a church a single blow and found that the reverberation from it continued for 30 minutes."

On a separate occasion, during a different NASA mission, a seismic test was performed where the Moon resonated for three days. After each of these events, there was speculation that the Moon is hollow. When examining this scenario, we need to look at the big picture. The Moon is enormous. The United States can

27

be placed on the near side of the Moon, as well as the far side. In between those two areas there is extra space. To go further into our analysis, we know that the Moon is approximately 2,158 miles in diameter. Inside of the Moon, it is believed that the Moon's thickness is over 500 miles. When we consider the math, it tells us that there is a 2000-mile diameter void inside the Moon. That is what is reverberating. Therefore, it was deduced that the Moon is hollow. The idea of the Moon being hollow fired the imaginations of some and perplexed others.

In 1979, two top government Soviet scientists by the names of Alexander Scherbakov and Michael Vasin put forward a startling theory. The two brought to the public's attention their hypothesis regarding the Moon's origin. They stated that our Moon is not a natural satellite, but an enormous spaceship created by extraterrestrials with highly advanced technology. They came to this supposition from data taken from the Apollo moon missions. Both men were reputable scientists that belonged to the distinguished Soviet Academy of Sciences. The two informed the public of their conclusions through an article that was presented in *Sputnik Magazine,* titled "Is the Moon the Creation of Alien Intelligence?" The article detailed their reasoning, as well as their conclusions about the Moon. From their research, the two concluded that it is very likely that the Moon is an ancient spaceship that traversed the universe and ended up in Earth's orbit.

They surmised that the Moon is a planetoid that was hollowed out eons ago, somewhere outside of our universe. They believed that these advanced beings used large devices to construct a massive spacecraft. They proposed that the Moon's originators used equipment to dissolve stone which created huge cavities in the interior of the Moon. Once this was accomplished, they believed that this enormous craft was navigated across the universe and positioned directly, and perfectly, into Earth's orbit. Shcherbakov and Vasin were quoted as saying, "It is more likely that what we have here is a very ancient spaceship, the interior of which was filled with fuel for the engines, materials and appliances for repair work, navigation instruments, observation equipment and all manner of machinery...in other words, everything necessary to enable this 'caravelle of the Universe' to serve as a Noah's Ark of intelligence, perhaps even as a home of a whole civilization envisaging a prolonged existence and long wanderings through space." From that time on, the "spaceship moon theory," was under discussion. It has become a consideration for how our Moon came to be here among Moon enthusiasts, ufologists, and researchers.

The Soviet scientists' assertion that the Moon is an ancient spacecraft, created outside of our universe by extraterrestrials, is about as outside of the box as one can get in determining the origins of the Moon. Vasin and Shcherbakov seem to concur with this sentiment stating, "Abandoning the traditional paths of 'common sense,' we have plunged into what may at first sight seem to be unbridled and irresponsible fantasy." However, the two believed that the Moon being a "spaceship," answered many of the baffling questions associated with it.

Vasin and Shcherbakov supported their idea of the Moon being a spaceship by closely examining some of the data from the NASA missions. This included the reverberations on the Moon which indicated that the Moon may be hollow. The two considered the fact that the lunar surface is two and a half miles thick, and that the Moon's craters, no matter what size, have approximately the same depth. The two scientists surmised that the Moon may have a purposely made, impenetrable hull. Scientists found in the samples brought back from the Moon that the lunar dust primarily consists of chromium, titanium and circonium. They suggested that if a material had to be fashioned to shield a "giant artificial satellite" from the effects of temperature, radiation, and meteor bombardment, scientists would likely have used these elements, because they are extremely resilient.

The Soviet scientists believed that this alleged ship had engines on the inside, but on the outside was coated with a moon looking substance to make it appear natural. On the spaceship moon theory, author and researcher David Hatcher Childress once commented, "The whole idea that our Moon is some gigantic hollow spaceship that's been put into a special orbit around our planet and contains cities and structures that are inside and outside of the Moon is to me a very reasonable assertion, and in fact it would seem to be that our Moon is some kind of gigantic artificial spaceship that is here to monitor our planet." The following are additional comments made by leading astronauts, astronomers, authors, NASA personnel, researchers, ufologists, and others regarding the Moon being a hollowed out, artificial object. We will keep their thoughts in mind as we move forward into the further information given in this book.

- "We cannot help but come to the conclusion that the Moon by rights ought not to be there. The fact that it is, is one of those strokes of luck almost too good to accept." —Isaac Asimov, Author

- "It seems much easier to explain the non-existence of the moon, than its existence."—Robin Brett, NASA Scientist

- "The more you study the Moon, the more you will become aware that it is an orb of mystery—a great luminous Cyclops that swings around the Earth as though it were keeping a celestial eye on human affairs." –Frank Edwards, Author and Researcher

- "It would seem that the Moon is more like a hollow than a homogeneous sphere." —Gordon MacDonald, NASA Scientist

- "They have no bloody clue where the moon came from and it shouldn't by physics be there." —David Icke, Author

- "The Moon is more complicated than anyone expected; it is not simply a kind of billiard ball frozen in space and time, as many scientists had believed. Few of the fundamental questions have been answered, but the Apollo rocks and recordings have spawned a score of mysteries, a few truly breath-stopping." —Robert Jastrow, Astronomer and Planetary Physicist

- "The Moon not only rang like a bell, but the whole Moon wobbled in such a precise way that it was almost as though it had gigantic hydraulic damper struts inside it." —Ken Johnson, former NASA Supervisor

- "How could we know the Moon is not a natural planet? The answer is that we know because it is too big, yet too light in mass; it is in the strangest position imaginable relative to the Sun and the Earth and it behaves in the most incredible way every month. What is more, it carries a whole series of quite specific messages that tell us it is an engineered' object." —Christopher Knight and Alan Butler, Authors

- "A natural satellite cannot be a hollow object." —Carl Sagan, Astronomer and Cosmologist

- "The origin and history of the Moon have remained a mystery despite intensive study by eminent scientists during the last century and a half." —Harold Urey, Nobel Prize winning Scientist

Art Imitating Life?

The idea of a cosmic entity being hollowed out and sent across the universe was around long before the Soviet scientists presented their summations about the Moon. In the popular 1960s television show *Star Trek: The Original Series,* one episode showed what it would be like to live inside of a hollowed-out sphere. In the episode titled "For the World is Hollow and I Have Touched the Sky" (third season, episode 8), the *Enterprise* crew races to prevent an asteroid from crashing into a Federation world, only to discover that the asteroid is really an inhabited, 200 miles in diameter spaceship. This asteroid-ship had all the characteristics of a typical asteroid. The creators of this camouflaged spaceship had programmed it on an autonomous trajectory through the galaxy. It adjusted for all gravitational stresses and ran on atomic power. On the outside, the surface appeared to be that of an asteroid, and even had a breathable atmosphere. Most interesting was the fact that when the landing party was transported onto the surface of the asteroid-ship, they immediately saw what appeared to be man-made structures, just as many have claimed of the Apollo astronauts when visiting the Moon. Additionally, the asteroid-ship was over ten thousand years old! The inhabitants of this pseudo asteroid were not aware that they were living in a hollowed out, intelligently constructed, artificial spaceship. What they were living on was not "only" an asteroid ship, but also a "*generational ship;*" people had lived out their entire lives on this artificial construction for thousands of years.

Some have proposed that our Moon just may be a generational ship as well. In the *Star Trek* episode, the people that inhabited this asteroid-ship referred to it as their "world." They didn't realize that it was a spaceship that had been traveling for ten thousand years on a course towards a new world. According to the story, the denizens of this craft were fleeing a world where their sun had gone supernova. The captain of the *Star Trek* starship *Enterprise* (James T. Kirk) surmised that some of the planet's inhabitants had been put aboard the ship and sent on a journey to another planet, which their descendants would reach thousands of years later, all the while preserving their race.

Could it be possible that this is the case with our Moon? Is art imitating life? We do not know what inspired *Star Trek's* creator, Gene Roddenberry, to produce an episode about a hollowed-out asteroid with a civilization living inside. Could he have had inside information about the Moon or some other cosmic entity, being an intelligently made, camouflaged spacecraft that was traversing the cosmos to a chosen destination? Roddenberry is said to have been sitting in on extraterrestrial channeling sessions. It is of my opinion that he received his inspiration for the *Star Trek* series

Gene Roddenberry.

32

during these sessions. One can only imagine what information he learned from those meetings. Perhaps some of the ideas for *Star Trek* shows began to formulate in his mind during that period. The episode titled "For the World is Hollow and I have Touched the Sky" certainly mirrors the idea of the hollow moon theory, and is nearly identical to the Soviet scientists' idea of a spaceship moon! It should be noted that this *Star Trek* episode and the scientists' research were created several years apart, with *Star Trek* coming first. It has been speculated that one of the ways that extraterrestrials are preparing us for first contact is through the media. If so, then Gene Roddenberry's *Star Trek* certainly seems to be accomplishing that task.

Two screenshots forom the Star Trek episode "For the world is Hollow and I have Touched the Sky."

Interestingly, the idea of the Moon being hollow and occupied, resembles another story, this time pertaining to the hollow Earth theory. In 1947 Rear Admiral Richard E. Byrd (1888-1957) allegedly flew his aircraft into an opening that led inside of the Earth. There, the admiral claims to have seen a gleaming city and interacted with members of a society living within a *hollow* Earth. According to the tale, this civilization is far more advanced than we surface dwellers. Additionally, he was given a warning to those living on the surface, stating that we need to smarten up, or we will obliterate everything on the planet. Byrd wrote down the events of this amazing journey in a diary. His experience has become legend. In fact, the theory that the Earth may be hollow has been discussed by various individuals and researchers for hundred of years. As far back as 1692, noted astronomer, Edmond Halley, considered the possibility of a hollow earth from a scientific point of view. Additionally, there is the idea that the Earth

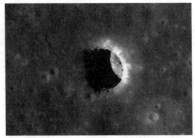

A Japanese lunar probe supposedly took this photo of a hole in the Moon.

may have openings at each pole where the hollow Earth can be accessed.

There are others that supported the idea of a hollow Moon as well, and they did so before the Apollo missions, or even that popular *Star Trek* episode. Noted British astronomer and selenographer Hugh Percy Wilkins (1896-1960) created the popular lunar observing guide *The Moon: A Complete Description of the Surface of the Moon,* that he co-wrote with astronomer Patrick Moore. Wilkins spent many years examining the Moon, which resulted in several excellent illustrations mapping the lunar surface. Wilkins believed that there were wide-ranging hollow sections in the interior of the Moon, possibly in the form of caverns. He believed that they were linked to the surface by large chasms. In fact, he located such an aperture within the Cassini A crater. Cassini A measures one and a half miles across, and the cavity that is heading down into the Moon's interior runs more than 600 feet diagonally.

In his book *Our Moon* Wilkins describes the interior of the hole stating, "Its inside is as smooth as glass with a deep pit or plughole, about 200 yards across at the center." In the 1920s Professor J.D. Bernal wrote *The World, the Flesh and the Devil.* He was of the opinion that humans sometime in the future would be traversing the cosmos in ships similar to a spaceship Moon type of vehicle; meaning that people would one day be dwelling in entirely enclosed inside-out-type worlds. In his book *J.D. Bernal: The Sage of Science,* author Andrew Brown wrote of Bernal's ideas for future space travel. Here we can see how closely his ideas matched the spaceship moon theory. Brown writes: "For man to live in space, Bernal argued, he would need to build a permanent, large sphere or 'celestial station'. The model he suggested involved a small spaceship attaching itself to an asteroid, which would then be hollowed out to provide a large, life-supporting, shell. He imagined it as an enormously complicated single-celled plant, with an outer protective wall that would be protective and rigid, as well as allowing the free access of radiant energy and preventing the escape of its internal atmosphere."

Today, researchers maintain that the Moon *could* be filled with enormous lava tubes and caverns. It is believed that many were formed around three to four billion years ago when the Moon had volcanic activity. As of late, lava tubes have been discussed by planetary geologists, astrobiologists, and surveyors as places to be considered as habitats for space settlements on the Moon, should we someday choose to colonize it. Researchers reason that lava tubes would be the ultimate set up for housing for future colonists. They would offer Moon residents fortification from such dangers as the ominous temperature fluctuations on the Moon, which can reach up to 120 C during the day and plummet to -130 C during the night. One can only imagine that if there are beings on the Moon, they too may have had the same idea, and may already be inhabiting these interior areas.

The notion that our Moon is not a natural satellite is difficult for many people to believe. The idea of some that it may be a spacecraft is even more problematic. Many still do not believe that UFOs are real; they will have a hard time trying to wrap their minds around the idea of the Moon being not only artificial but an inhabited, alien spacecraft! This concept is instantly compared to the Death Star in the *Star Wars* movies and dismissed as fiction. However, those that have long been a part of the research into the Moon recognize that this may not be as farfetched as it sounds. Alexander Scherbakov and Michael Vasin put their reputations on the line by publishing their theory. Both had a lot to lose coming forward as some scientists may have disputed their claims. However, they felt this information and their deductions were important enough to move forward with their hypotheses, whether they were believed or not. There are several theoretical scenarios that can be attributed to the spaceship moon supposition.

A photo of the Moon from Apollo 16. (NASA)

Hypothetical Scenarios for a Hollow Spaceship Moon

An Escape Ship

The first hypothetical spaceship moon scenario can be likened to Noah's Ark of biblical fame. If the Moon really is a gigantic spaceship, we can only imagine the reasons that a race of beings would embark on a project of such a grand scale, as it takes light years to traverse the universe. It could very well be that they were trying to avoid a disaster. Our Moon could just be a lifesaving spaceship where extraterrestrials escaped a dangerous situation on their home world, and literally fled for their lives. We can only speculate what type of disaster, but certainly one large enough to cause them to build a ship to escape. We can image too that it was not a quick ending. They may have known for a long period of time of the pending disaster, and thus were able to build what we are now calling a "spaceship" of epic proportions.

It could also be a "generational ship," with the beings still living there today. We could spend a great deal of time imagining the different scenarios that would cause a race of people to leave their home world. In the biblical story of Noah's Ark, God warned Noah of the pending doom of a worldwide flood far in advance and directed him to build a ship in which a few of the populace were placed along with animals. Then when the great destruction came in the form of rain, Noah and the chosen ones went inside and closed the door. For extraterrestrials that may be existing on the Moon, it could have been a similar situation, where there was a war on their world or some sort of natural catastrophic disaster, or even an impending asteroid that could not be avoided. Or as in the *Star Trek* episode mentioned earlier, it could have been a dying sun going supernova.

If this were an escape situation, it would be prudent to consider whether the inhabitants left spaceship Moon and transported to Earth once they reached our orbit. Could this group of beings have been fleeing for their lives and in their efforts to preserve their race, parked their ship in Earth's vicinity? Could they have brought a DNA bank with them to further preserve their species and other life on their doomed world? Maybe they even split into two groups, sending some to Earth while the others remained on the craft, in orbit around Earth. If this were the case, could those that remained on the Moon have done so to keep watch over their people on Earth? Could these same beings be watching us from the Moon today? If so, have they been traveling back and forth from the Moon to Earth this entire time? Do they have operatives on Earth? This could explain why the Moon gives the impression of being a reconnaissance satellite. Perhaps they are overseeing their people's progress here on Earth.

City Ship

A transportable city is an idea that can be found in today's science fiction. However, given what we have just covered about the possible origin of the Moon, it is within reason to consider it from this perspective. Could it be that the Moon itself is an enormous spherical city ship, created by an extraterrestrial race? A city ship would essentially be a traveling world. In this scenario, the inhabitants of this ship are not fleeing, but are purposely traveling the universe. In recent years with the advancement of telescopes and cameras, people have captured astonishing videos of the Moon. Several of them reveal what appears to be spherical shaped objects flying across the Moon and entering and exiting large chasms. It appears that the Moon is harboring small spacecraft inside of it. Are these explorers? Are they scientists? Are they using their place in Earth's orbit as a holding spot for their large ship, while the smaller ones travel the galaxy? There have been those that claim they have witnessed water being extracted from rivers and other large bodies of water on Earth. Could it be that if there are beings on the Moon, they are using some of Earth's resources? Might they one day decide to navigate this potential great ship into another system once they are done with ours?

A Garden of Eden

In the Garden of Eden hypothesis, the Moon was transported into Earth's orbit by advanced beings to aid and support Earth and all creatures on it. Their aim would be to create a global paradise where life would flourish. In this scenario, we consider whether they brought seeds of plant and animal life from their home world and deposited them on Earth as well. Some researchers believe that the Earth was partially terraformed. Others maintain that humans did not originate here. Still others believe that humans are a hybrid race that are a combination of primates and extraterrestrials. There is also a blood type in some humans that scientists cannot explain (RH Negative). Could some of the populace in a spaceship Moon scenario have come to Earth, lived among us, and bred with humans? There are tales in certain religions and myths around the world that speak of such unions, referring variously to these beings as "angels" and "gods." Is this who we really are...the descendants from a race of beings who thousands of years ago mixed with humans? We can only speculate if we are their offspring and if this is where our origins began. *Could they be us?*

A Mechanism

One theory states that the Moon is some sort of mechanism or piece of equipment that is used by extraterrestrials to monitor Earth and other areas of the Solar System and report back to higher beings that are seeking this information. In this scenario, the Moon has also been referred to as a machine. The beings believed to be on the Moon may be the keepers of this information, and the caretakers of the Moon. Sources state that the Moon is filled with technology to collect this information and has been doing so for millions of years. It is thought that at some point there were beings that became aware of the Earth and sent in this large piece of equipment (the Moon) to engage in what can only be called a lookout station for what is going on with Earth and her inhabitants. This same group of beings is more than likely observing other areas, beings, and worlds as well.

A Space Station

Some have likened the Moon to a space station, often thinking of the *Star Wars* movie in which character Obi-Wan Kenobi utters, "That's no Moon, that's a space station!" This megastructure is believed to be a gigantic, artificial satellite designed to remain in orbit around Earth for a long, possibly indefinite period. A space station is generally thought to include workstations and a living environment. The station would be divided into separate compartments, rooms, or small houses. It would include life sustaining facilities, and be able to support a self-contained community.

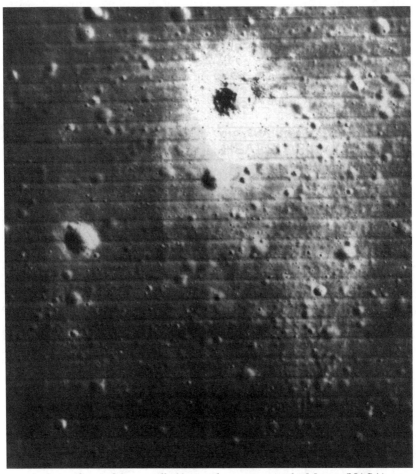

A photo of one of the so-called bottomless craters on the Moon. (NASA)

Chapter Three

An Inhabited Alien World

"The Universe is a pretty big place. If it's just us,
seems like an awful waste of space."
—Carl Sagan

Two thousand, five hundred years ago in ancient Greece, the philosopher Anaxagoras (ca. 500-480 BC) began his pursuit of studying the Moon. He was eventually exiled for teaching that the Moon is not a deity, but a rock. Of course, he was correct in his position that the Moon reflects the light from the sun, which allowed him to explain the Moon's phases as well as the eclipses. The Greek philosopher Plutarch (ca. 45-125) wrote about the "face appearing in the rounded orb of the Moon." The invention of the telescope played an important role in advancing our understanding of the Moon. After learning of the "Danish perspective glass" in 1609, Galileo Galilei (1564-1642) created his own telescope. Galileo lived during a period when many of the prominent philosophers closely adhered to Aristotle's teachings. They believed that anyone who disputed Aristotelian philosophy was wrong.

However, Galileo enhanced the telescope he developed into one that was able to increase magnification up to thirty times. He began to observe the Moon and proceeded to illustrate his findings. Galileo observed that the lunar surface is not smooth, uniform, and precisely spherical as many of the philosophers believed, but found that it is uneven, rough, and full of cavities and prominences, very much like Earth. In March of 1610, Galileo published his early findings of his telescopic observations of the Moon (as well as those of the Sun, Jupiter, and the phases of Venus) in a periodical titled *Starry Messenger (Sidereus Nuncius)*.

His work quickly became popular, as people were eager to have a better understanding of the cosmos. Galileo's illustrations of his observations gave people a different idea about the Moon. Readers soon understood that what they were seeing in Galileo's views of the Moon was very different from what they had previously learned. Galileo was unique in that he had studied Renaissance art and understood chiaroscuro (a method used for shading light and dark). It was therefore easy for him to discern that the shadows he was witnessing on the Moon were mountains and craters. From his drawings, he was then able to approximate the elevations and depths. His discoveries about the Moon showed that Aristotle's idea of the Moon being a perfect orb was incorrect, and that in fact, it was like that of Earth in its topology. However, this information was not widely accepted, as most continued in their belief in Aristotle's teachings.

In those early days of the telescope, there was the belief that the Moon was an alien world, complete with a populace of other worldly beings and cities. In fact, from the time the telescope was invented, and for centuries afterwards, it was considered scientifically correct by several scientists. German astronomer Johannes Kepler (1571-1630) started the search for lunar people in 1610. Prominent astronomer William Herschel (1738-1822) maintained that we are not alone in the universe, and regularly sought out signs of life on the Moon. He determined that there was a civilization of beings residing on the Moon and dwelling within its craters. He once stated, "Who can say that it is not extremely probable, nay beyond doubt, that there must be inhabitants on the Moon of some kind or another?" Joseph Smith (1805-1844), the founder of Mormonism, believed in lunar inhabitants, stating, "The moon is inhabited by men and women the same as on the Earth, and they live for a greater age than we do—that they live generally to near the age of 1,000 years." Other believers included the well-regarded German philosopher and astronomer Nicholas of Cusa (1401-1464), and German astronomer Johann Hieronymus Schroeter (1745–1816). Schroeter recorded changes in the Linne crater. He created several diagrams depicting the Moon that included the six-mile crater over the years. Over time, he observed that Linne's size was decreasing. Today Linne is but a small hole with barely any height or depth. Schroeter believed that the dwindling of Linne was the result of work by "Selenites" (the name during that period for lunar inhabitants). Investigative journalist and author Philip Coppens, in an article for *Frontier Magazine* (1995) titled "The Alternative Conquest of the Moon" writes, "As early as 1788 Schroeter had observed small 'swollen parts' on the Moon. He argued that these were the result of industrial activity of the 'Selenites,' the inhabitants of the Moon."

American astronomer William Henry Pickering (1858–1938) made numerous noteworthy Moon observations. On one occasion he believed that he had discovered insects on the Moon. He was surprised one day as he studied the Moon's craters (especially the Eratosthenes crater) to see what he described as "traveling dark objects" that were moving across the surface of the Moon. He wrote of this strange sighting stating, "In trying to find conclusive arguments for or against the existence of animal life upon the Moon, I have necessarily studied not only the routes along which it appears to travel, but also the reasons for which it might be expected to travel." Pickering surmised that what he had seen that day were a form of Moon insect. He reported that they covered 20 miles in 12 days. Pickering believed that these "insects" were the cause of the changes that he found in the Eratosthenes crater. In his book, *Our Mysterious Spaceship Moon* researcher Don Wilson writes, "no one questions Pickering's integrity and competence. He did see something. What it was remains a mystery."

Nicolas Camille Flammarion (1842–1925) was a renowned astronomer from France and the author of several popular science fiction novels. He was also the owner of *L'Astronomie* magazine, and had a personal observatory in the city of Juvisy-sur-Orge, France. He proposed that nearly all the planets in our Solar System are inhabited. He considered the Moon to be a populated world, complete with an atmosphere, water, and plant life. It was Flammarion who coined the term "Selenites" for lunar inhabitants. Most interestingly, when examining Flammarion's research on the Moon, some have questioned whether he had observed something entirely different from what we see there today. We can only speculate whether he had seen things even more enigmatic than the strange events on the Moon today, and if they have since vanished.

The idea of an inhabited Moon has persisted up until today. In examining the astronomers of the past and hearing their views about the Moon possibly being populated, one can only wonder if we have come full circle. Could the astronomers of those early years have been correct all along? Is the Moon inhabited?

Is Someone There?

When we look at the information gathered from the long line of researchers, astronomers, scientists, astronauts and others, there is no denying that there quite possibly *is* someone on the Moon. The odds of so many reputable people being mistaken about there being unexplained events and strange phenomena occurring on the Moon are extremely low. The following are just a few of the comments put forward by people that have been investigating the Moon and its strange happenings for years. Ukrainian Radio Astronomer and Astrophysicist Dr. Alexey V. Arkhipov once stated, "Our Moon is a potential indicator of a possible alien presence near the Earth at some time during the past 4 billion years." NASA Astronaut Scott Carpenter stated, "At no time, when the astronauts were in space were they alone: there was a constant surveillance by UFOs." American Biologist and Writer Dr. Ivan Sanderson commented, "Many phenomena observed on the lunar surface appear to have been devised by intelligent beings." Popular French Historian and Author Robert Charroux writes, "Are we to conclude that these lunar craters have been frequented by extraterrestrial astronauts? The possibility cannot be rejected, especially regarding the Plato crater, where many mysterious lights have been observed." Former Marine Air Corps Major Donald Keyhoe states, "All the evidence suggests not only the existence of a Moon base, but that operations by an intelligent race have already begun. If so, who could the creatures be? Were they from other planets or did they originate on the Moon?"

The truth is, the more we learn about the Moon, the stranger the Moon's story becomes. One thing that is apparent, is that the citizens of Earth do not own the Moon. It appears that somebody else does. Some may find the whole idea to be ludicrous. However, we cannot deny that there is something extraordinary going on with the Moon, and it is *not* coming from us. No matter how we spin this, there is no clear explanation for all the strange, perplexing, anomalous activity on the Moon. There are numerous stories of UFOs on the Moon, weird sightings by astronauts on the way to the Moon and while working on the lunar surface—strange lights, movement, and more. If just one of those stories is true, then that means that we are not alone in the universe, and that someone else is on the Moon. In addition, if we find that there are extraterrestrials inhabiting the Moon, then there is a likelihood that they have been there for a long period of time. They may have been there as far back as our ancient times.

There are petroglyphs and other ancient artwork of what appears to be strange aircraft and beings wearing space helmets. There are tales going back centuries from around the world of visitors that came from the stars to Earth. Given that the Moon is near Earth, there is a chance that the beings portrayed in images from the past, as well as in the stories, came from the Moon.

The Moon just may be the key to *learning the origin of mankind, *mankind's possible connection to the tales of extraterrestrial visitors in ancient times, *the validity of a worldwide flood, *whether or not mankind was a part of a civilization that started over, *if we have already had first contact, *Earth's connection to the Moon and Mars, *proving whether or not we are alone in the universe, *where the UFOs seen on Earth are coming from, *where the alleged extraterrestrial visitors on Earth came from in the past and more. It is possible that the Moon is the key to our many questions. However, the main question that people want answered is whether we are alone in the universe. Once this question is answered, mankind will be able to move forward and enter an era unlike any other in human history. What sorts of scenarios might an inhabited Moon be engaged in? The following are several theories for an inhabited Moon.

Inhabited Moon Theories

Foreign Country Theory
Could the Moon be an established alien country or nation? We should consider the hypothesis that there is an established civilization there with its own set of laws and government, where

the beings consider themselves and the Moon completely separate from Earth. The idea that the Moon is the equivalent of a foreign country with a civilization dates back to ancient times. There were many philosophers in the past who believed that there was life on the Moon. Among these were Anaxagoras of Clazomenae, Aristotle, Lucian of Samosata, Plutarch, Pythagoras, and Xenophanes. The noted Natural Philosopher and Author John Wilkins in the mid 1600s expressed his idea of there being life on the Moon in his work titled *Discovery of a New World in the Moon,* stating, "It is probable there may be inhabitants in this other World, but of what kind they are is uncertain." In more recent times, people have also espoused the idea of there being extraterrestrial life on the Moon. Some believe that these lunar people are the descendants of advanced beings that originally brought the Moon into our Solar System, purposely placing it in Earth's orbit. In an article written by researcher Steve Omar titled "Moon and Mars: The Moon is a Foreign Nation," Omar pondered, "Could it be that the Moon is a foreign country and someone else's property, and the Moon's government does not want us coming up and invading their territory with our nuclear weapons, pollution, unwelcome military facilities, diseases, litter, mining exploitation, and historically proven record of foreign imperialism?"

Is it possible that what we have heard for centuries from philosophers and astronomers regarding the Moon being populated been correct all along? Are there Moon aborigines? Our science tells us that life cannot survive on the lunar surface. However, could we be missing something? Could there have been a time in the Moon's history when there were conditions where life could have survived even for a short period of time on the surface? If there were lunar natives originally, we might consider the possibility that they moved underground due to some disaster on the Moon. Perhaps these potential aborigines went inside of the Moon eons ago and have been there ever since. Researchers have found what they believe are ruins from structures on the Moon. These alleged ruins could be from a time in the Moon's ancient past.

What's interesting about this scenario is that it helps us to recognize that the Moon does not belong to Earth. It just may belong to otherworldly beings that consider it home, with their own culture. It is possible that if this is the case, they may not want to have anything to do with Earth and its inhabitants. Could it be that if there are extraterrestrials on the Moon, they are watching our progress to make certain that we do not infiltrate their world? Is it possible that when objects were crashed into the Moon by NASA, there was someone there that felt the reverberating? Did they see the astronauts land on the Moon? Were they nearby observing them at work? If the Moon is a foreign country, then perhaps the rumors are true, and we really were "warned off the Moon." It could very well be that there are lunar inhabitants, and just as the nations of Earth do not allow "aliens" from other countries to enter their land without permission, perhaps lunar citizens are the same about the Moon. It just could be that an extraterrestrial populace on the Moon does not want their neighbors from Earth intruding on their homeland. It just may be that when we travelled to the Moon, that *we* were the *aliens* intruding upon *their* world.

Colony Planet Theory

The more we learn about the Moon, the better we understand that we are dealing with something that is far more than "Earth's Moon." Although the Moon is Earth's satellite, it is not necessarily a moon according to the traditional sense. We have already established that the Moon behaves differently than any other moon in the Solar System. Just as humans have the idea of colonizing the Moon, this could have been done already by other beings searching for a home or seeking to expand their territory. They may have been there for millions of years already. In the 1950s and 1960s, a number of movies came out with the theme of people landing on the Moon with the goal of eventually colonizing it. Could this have been a message to the public as to what is really going on with the Moon, and to prepare people for the news that someone is already there? Or was it to prepare us for what the future may hold regarding space exploration and what we can expect? We also do not know why an extraterrestrial race would choose an otherwise difficult world to establish and grow a colony. Perhaps when humans find themselves in the same position, we will realize the benefit of inhabiting a planetary object with a not-so-wonderful environment. It looks as if our time is coming. More will be covered in Chapter Ten, *Into the Future*.

Stopover Point Theory

It has been hypothesized that UFOs are using the Moon as a stopover point either as they travel through space or when Earth is their destination. It is also thought that this has been going on for millennia. This is especially relevant since there are a variety of UFOs seen around the Earth in various shapes and sizes, with the more prominent ones being saucer shaped and elliptical.

British journalist and UFO researcher Harold T. Wilkins, in his book *Flying Saucers from the Moon*, states, "I have suggested that the moon may be and long has been a stopover place for what we call flying saucers, or spaceships." In *Flying Saucers on the Attack*, he mentions it again stating, "It was near the close of the 18th century when several men in London and Norwich saw strange lights on the moon that appear to indicate that our satellite was being used as a stopover place, in flights of observation to the Earth."

If Wilkins' assumption is correct, that may explain the mystery of UFOs in Earth's vicinity, and the many different types of alleged mysterious spacecrafts that have been reported. Some will find this hypothesis farfetched given how long it takes to travel across space, which can amount to thousands of years. Some will scoff and say that it is unlikely to see one UFO, yet alone many different types with so many that the Moon would be a stopover point. Still the stories of UFOs around the Moon and Earth persists and have increased over the years. Therefore, it is safe to place this in a hypothetical situation and one that we can attempt to decipher.

We do not yet know or understand the advanced technologies that extraterrestrials may hold. Travel through wormholes has been theorized to provide possible short cuts through space, therefore allowing a variety of extraterrestrials to come through, using the Moon as a stopover point before visiting Earth and other planets in the Solar System. This mode of travel is often seen in science fiction. We should ask ourselves, how often has science fiction morphed into real life? The answer is more often than we can keep track of. Case in point, again during the 1950 and 1960s there were several movies made about traveling to the Moon and encountering extraterrestrials long before the Apollo missions were established.

The stopover point hypothesis makes sense when we see the vast number of UFOs around Earth. This is no longer a phenomenon that people can ignore. Too many people have seen them and documented them in photographs and videos. At the same time, government officials are now talking publicly about them. So then, where are they coming from? The Moon is a good place to consider. If the Moon is indeed a stopover point, then these UFOs seen around the Moon and Earth could be from a variety of extraterrestrial groups. Of course, the reasons for their visitations are unknown. According to ufologists and researchers, theoretically speaking, extraterrestrials may be using the Moon for a variety of reasons including, bartering, mining, resting, even repair and refueling of spacecraft. Once they leave the Moon and head towards Earth, their reasons may vary as well. They could be visiting Earth out of curiosity, for scientific work, exploration, Earth's resources—we can only speculate. There are numerous stories out there of people that claim to have had extraterrestrial contact. There are thousands of reports of people witnessing UFOs in both the day and night skies. It would behoove us to investigate these claims and consider whether they are connected to the Moon.

Observation Post Theory

There is a theory that extraterrestrials have been watching mankind since the beginning of human history, and that the Moon is their "observation post." Additionally, if true, and extraterrestrials are on the Moon watching, some believe that they are particularly interested in our space ambitions and advancements. It is believed also that the NASA programs including those of Mercury, Gemini as well as the Apollo missions were all watched and monitored. Many were even followed by UFOs while in flight. In his book *Our Cosmic Ancestors*, former NASA Communications Engineer Maurice Chatelain writes, "Moments before Armstrong stepped down the ladder to set foot on the Moon two UFOs hovered overhead."

In his book *Visitors from Other Worlds*, Author and Researcher Brad Steiger writes about Dr. Sergei Bozhich who witnessed the

Russians monitoring the *Apollo 11* Moon landing. Steiger writes, "In his [Bozhich's] opinion, the two UFOs appeared ready to assist the US astronauts in case anything should go wrong with the landing. Once the module appeared to be securely settled on the lunar surface, the alien spacecraft flew away." These accounts indicate that if indeed extraterrestrials are watching mankind, the Moon is where they are observing from, as well as the possibility of sending craft from there to Earth. This could explain the many sightings of UFOs seen around Earth and those reported near the Moon.

Mining Post Theory

Another scenario is that of there being a mining post on the Moon. It is the idea that extraterrestrials are mining for resources.

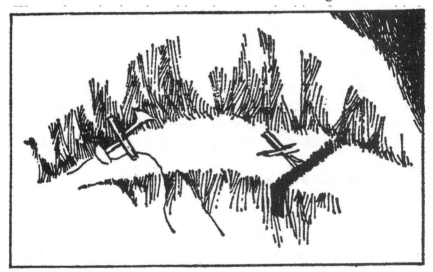

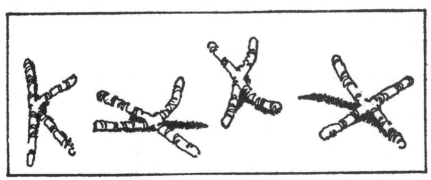

George Leonard's concept of the X-drones that are used to mine the Moon.

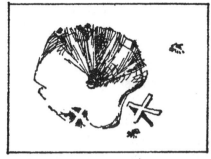 

George Leonard's concept of the X-drones and spray from a Moon crater.

the Moon. The goal of this alleged civilization is to have people living off Earth in the event of a worldwide catastrophe where the human race could not be saved. Another speculation as to why there would be a breakaway civilization on the Moon is to be able to discern whether humans can live comfortably on other worlds than Earth. In this case, the breakaway civilization would learn what it takes to live on another planet and even possibly with extraterrestrials. The Moon and Mars have been suggested as off world sites for such an experiment. The people in these colonies would be observed to perceive how they performed outside of Earth in an alien environment. In the case of the Moon, it is thought by some that a breakaway civilization may be operating from the Moon's interior as well as parts of the surface. There are several ways that people could become part of a breakaway civilization. It is thought that individuals would be selected for their field of expertise. These would be areas helpful in creating and operating a settlement. Others could volunteer or be recruited. All would be young and extensively trained.

The Duck Blind Theory

Could the Moon be a type of "duck blind?" On Earth, a duck blind is a shelter, camouflaged with foliage for concealing duck hunters. A disguised satellite would give extraterrestrials the means to observe mankind at close range. They would be hidden and thus would not interfere with the people of Earth's development. The Moon may quite simply be a duck blind camouflaged to appear as a Moon to those on Earth (or elsewhere). It could be designed to look like a planetary body, all the while being home to extraterrestrials that are watching and observing Earth. From all the theories that I have put forward in this book, this one shockingly stands out.

Science Vessel Theory

We should consider whether the Moon was initially a type of science vessel. Could the Moon have been sent here on an alien scientific research mission or one of general exploration? Could it be that extraterrestrial races in our universe beat us to the *Star Trek* fantasy and are already out there, "seeking out strange new worlds?" It is possible that one of those "new worlds" could have been the Earth? There may be races of beings out there that are studying our world, but who do not make themselves known and are simply collecting data, with no intention of connecting with us in any way except for scientific means.

Extraterrestrial Defense Satellite Theory

An extraterrestrial defense satellite is the idea that the Moon houses beings set up as sentries to protect the Earth from malevolent extraterrestrials and other dangers. Could this be the reason that our astronauts have been safe in space and on the Moon despite the reported UFOs they encountered along the way? Could they have been keeping their distance because they were warned to leave the astronauts alone? On the other hand, could the UFOs that the astronauts were followed by in space be guardians? Perhaps they were not followed merely out of curiosity as some believe. They could have been acting as sentinels protecting the men within the ships. Could there be benevolent and malevolent extraterrestrials out there with an interest in Earth?

Might there be benevolent beings that feel that because we are not technologically advanced enough to defend ourselves from nefarious alien beings, that we need protection? Could it be that there are beings in the universe that have protected us and prevented a takeover of Earth? Better yet, could the rumors of a "Galactic Federation" be true and we fall under their jurisdiction, and we are protected without even realizing it? There are reports that a couple of the Apollo missions ran into trouble in space, and some felt that they may have been assisted by extraterrestrials. Could the extraterrestrials be acting as protectors over humanity? Could there be a star wars waging over ownership of Earth that we know nothing of, and there are benevolent factions on the Moon? There are also accounts of weaponry being mysteriously disarmed on Earth, as well as asteroids being moved out of the path before causing a disaster. Some believe that this is the work of extraterrestrials. If so, are these beings operating from the Moon?

Bases on the Moon Theory

There have long been rumors that there is a secret base on the Moon placed there and operated by some Earth governments. This hypothetical base is believed to be filled with personnel that were trained and sent to work on the Moon. Some believe that this base is operated only by humans. Others contend that humans and extraterrestrials are collaborating on the Moon. In 1994, NASA's *Clementine* space probe took pictures of what some believe is an artificial construction. The figure is shaped like a V and is located on the far side of the Moon, in Mare Moscoviense. It appears to be made up of two rows of seven lights. The design is symmetrical and measures roughly 500 ft by 420 ft. To some researchers this formation resembles a base, and they consider it proof that a Moon base does in fact exist.

In his book *Nothing in This Book Is True, But It's Exactly How Things Are,* (page 179), author Bob Frissell talks about a secret organization that has established a base on the Moon. He writes, "First they made a base on the Moon, using it as a satellite to go deeper into space. They built three small bubble type cities on the dark side. There was an accident on one of these and many people were killed. Records will indicate that there have been more than 2,000 secret missions to the Moon."

In speaking strictly about an extraterrestrial base, lunar researchers and ufologists have speculated for years about the possibility of there being an extraterrestrial base on the Moon. Americans have dubbed the base Luna. *The Encyclopedia of Moon Mysteries, Secrets, Conspiracy Theories, Extraterrestrials and More*, tells us that Luna is "the name given to an alleged extraterrestrial base located on the far side of the Moon. Purportedly, the Luna base has facilities and a mining operation. It is also an area where extraterrestrial spacecraft are believed by some, to be parked." From where was this information obtained?

George Leonard's concept of the bridges and stitching on the Moon.

The Encyclopedia of Moon Mysteries under the topic of "bases" continues, "There is a story of a NASA employee that discovered a base in Apollo photographs in the 1960s. Reportedly, the pictures showed an extensive complex that had structures in a variety of shapes and sizes; including geometric shapes, towers, as well as mushroom and spherical shaped buildings." Researcher and author Harold T. Wilkins writes in his book *Flying Saucers on the Attack*, "I advance the theory that our moon has been, and still probably is used as an advanced observation base, in regard to our Earth, by mysterious cosmic visitants connected with the flying saucer phenomena."

There are there theories as to why extraterrestrials would establish a base on the Moon. 1) *To Observe Human Progress.* The theory that extraterrestrials are monitoring the development of life on Earth. A base on the far side could be easily hidden from humans. 2) *To Monitor Usage of Nuclear Bombs.* In the 1950s, when the atomic bomb was first created, there was an increase in UFO sightings. During that same period, extraterrestrials are said to have come to the Earth and visited certain governments in an effort to stop the usage of this deadly weapon. 3) *Mining of the Moon.* One prominent theory is that extraterrestrials are mining the Moon for materials, which would require a base of operations. Some speculate that if indeed there are extraterrestrials with bases on the Moon, they could be a group of beings from outside of our Solar System.

There is also the suggestion that there may be several extraterrestrial groups working there, each with its own base with different agendas, or with the same goal and collaborating together. A mysterious tale states that allegedly the Germans established a base on the far side of the Moon in the 1940s, years before the United States placed men on the Moon. They purportedly had advanced technology to help obtain this goal. It is said that they obtained knowledge from recovered UFOs that had crashed on Earth. Others contend that the Germans had contact with extraterrestrials and information was obtained directly from them. Once

George Leonard's drawing of the huge structure inside a bottomless crater on the Moon.

they established a base on the surface, they allegedly began tunneling underground. In his book, *The Watchers: Who Watches Us from the Moon?* (pages 15-16), Roger King writes, "Milton Cooper, a Naval Intelligence Officer tells us that not only does the Alien Moon Base exist, but the U.S. Naval Intelligence Community refers to the Alien Moon Base as 'Luna,' that there is a huge mining operation going on there, and that is where the aliens keep their huge mother ships while the trips to Earth are made in smaller 'flying saucers.'"

Who Owns the Moon?

In 1969 NASA launched the Apollo missions to gain information about the Moon. Several scientific experiments took place, and mirrors and seismometers were placed there for us to collect data about the Moon. The astronauts were to examine the Moon for the possibility of one day placing colonists there. In 1994, the Ballistic Missile Defense Organization (U.S. Navy) and NASA dispatched the *Clementine* spacecraft to chart and take images of the lunar floor. Both the Apollo missions and that of *Clementine* resulted in photographs with Moon anomalies. There was photographic evidence of there once being intelligent life on the Moon. According to sources, much of the evidence was changed, hidden, or placed away from the public's eyes. Today, we have several pictures that reveal intelligently made objects on the Moon. We have the testimony from astronauts and others about there being life on the Moon. We also have modern day astronomers that are consistently posting their findings of objects on the Moon, and they are startling! There has also been talk of the face of the Moon changing and the belief that extraterrestrials are responsible for it.

The interesting question in all of this is, "Who owns the Moon?" Does the Moon belong to Earth since it is fundamental to life here? Or if there are lunar inhabitants (especially if they have been there since our ancient times) do the extraterrestrials own it? If our governments were to determine that the Moon belongs to Earth, then who do the possible Moon cities, and anything found on the Moon belong to? Even more interesting is the idea that the ancient artifacts located on the Moon were abandoned eons ago. There could be new inhabitants on the Moon now. Do they own what is on the Moon or does Earth? We also need to know who the original beings were. Were they connected to us and why were they on the Moon in the first place? Even more interesting is the question of, "*Are we them*?" Could the Moon inhabitants have originally come from Earth? If so, did they travel there in ancient times, or are they time travelers as some have suggested? If Earth had an ancient past as suggested earlier, did humans use flying machines to go to the Moon and establish a civilization there?

If there is any truth to extraterrestrials visiting the Earth as far back as our ancient times, and even prehistory, if there is any connection between us and lunar inhabitants travelling between the two worlds, and even having influences on humans as stated earlier, then they may have continued to visit humans through the ages, even until now. If there were visitors from the Moon, they were likely here throughout humankind's history. These may well be the same beings whose ships were depicted in petroglyphs, cave art, ruins, medieval art, renaissance paintings and other artwork that researchers believe to be beings from other worlds. They just may be the same extraterrestrials whose stories have been told about the gods of old, through religions and traditions known today.

In my studies of extraterrestrials, the UFO phenomenon, and the unseen world (i.e., life after death, other dimensions, etc.), I have come across sources that indicate that the Earth is highly regarded and special. It is considered unique among those from other worlds and even other dimensions. It is thought to be the *jewel* of the Solar System. Therefore, could Earth be a kind of tourist attraction to the cosmos? Are some visiting the Moon on their way to Earth, using it as a stopover before arriving here? There is also information that suggests that some of the life on Earth was brought here from other planets. Could there have been several groups of beings that terraformed the Earth, and now use the Moon as a stopover point or a space station in their monitoring of mankind and other lifeforms on Earth?

One wonders if the beings on the Moon may have been overseeing who visits Earth all along, with the Moon being the watchful eye on the survival of this unique jewel that we live on and a center to monitor who visits. There are many stories of extraterrestrials visiting Earth going back ages. We know of the gods of old, but there are other tales of spacefaring beings coming here throughout time. For example, in the state of Utah, there are petroglyphs that portray what appear to be beings dressed in spacesuits resembling what our astronauts wore in the Apollo missions. There is talk of there being humans on the Moon that are time travelers. Could the imagery in the petroglyphs have any connection with the Moon and astronauts from the future? *Does the Moon hold the answers to our past?*

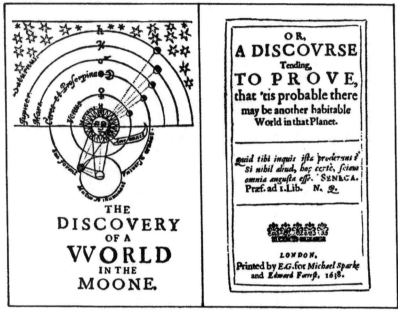

A 1638 book called *The Discovery of a World in the Moone*.

A 1687 book on the Moon by Cyrano Bergerac with a frontispiece showing him
ascending to the Moon in a glass ball.

Chapter Four

Ancient Aliens and the Earth Moon Connection

"The all but impossible glory of having walked on the
moon, of proving our mind power and our brilliant
technology, this cannot ever be dimmed... We have
hurled ourselves closer to the Gods."
—Emil Petaja

Does Earth have an ancient alien connection to the Moon? Is the
Moon a part of Earth's galactic history? The term "ancient aliens"
has become popular in recent years. Also referred to as "ancient
astronauts," it refers to advanced beings that theoretically came to
Earth and interacted with primitive humans during ancient times.
Ancient astronaut proponents maintain that this interaction with
humans was no accident. It is thought that these spacefaring beings
came to help humans get a footing in this world by instructing them
on needed knowledge and information to advance themselves
and progress. They taught mankind the ways of survival. As a
result, humanity thrived and continues doing so today. In those
early times, the extraterrestrials expanded humanity's worldview
by giving them advanced knowledge and understanding that
aided them in establishing cultured civilizations and giving them
the basic knowledge and skills of survival allowing humans to
eventually flourish on their own. They are the ones that lit the
spark in humans establishing themselves. In time, they left
humanity to its own devices. Now it appears that they may be
waiting for mankind to advance enough for them to invite us into
what just may be a galactic community. It is thought that primitive
humans embraced these beings as gods due to their knowledge

and most especially their technology. We can only imagine their reaction as they flew in from the sky and landed spacecraft on Earth. In an article titled "Forbidden History," for the *Unariun Wisdom* website, researcher David Icke explains, "Contrary to conditioned belief, life on Earth has not evolved from a primitive past to the technological 'cutting edge' of today. Many thousands of years ago, as detailed in streams of ancient accounts across the world, there was great technological knowledge on this planet and a global society controlled by races of beings, which humans came to know as 'gods.'"

Centuries later, evidence of a civilization would be found on the Moon by US probes and manned missions. This evidence includes what appear to be ruins of an ancient city. One wonders if these ruins may have a connection to the beings that assisted humans in ancient times. Were they the same beings that were mistaken for gods? When these ruins were first discovered on the Moon, it was such a shock that the information was not shared with the public. Until this day, there has been no open admission to any of this from the National Aeronautics and Space Administration (NASA), even though there is a good amount of evidence by way of pictures, NASA personnel coming forward with stories, and constant activity seen on the Moon by astronomers for centuries.

After the Apollo missions, several astronomers and researchers wrote books about the discoveries on the Moon. Some of the most notable were *Our Mysteries Spaceship* Moon (1976) and *Secrets of our Spaceship* Moon (1979) both by ufologist and author Don Wilson. George Leonard's *Somebody Else is on the Moon* (1977) gave even more detailed information about structures on the Moon and more. In 2006 Christopher Knight and Alan Butler came forward with their book *Who Built the Moon*, which was even more eye-opening as they promoted the theory that the Moon is not a natural satellite.

In 1996 a meeting was held at the Washington National Press Club by lunar researchers intent on informing the public about the mysterious discoveries located on the Moon. An excerpt from the official press release states: "NASA scientists and engineers participating in exploration of Mars and Moon reported results of their discoveries at a briefing at the Washington National Press

Club on March 21, 1996. It was announced for the first time that man-caused structures and objects had been discovered on the Moon." Some of the information presented that day included: *The Soviet Union having photographs that could prove the presence of extraterrestrial occupation on the Moon. *Numerous photographs revealing several areas of the lunar surface where evidence of extraterrestrial activity was apparent. *Photographs and film footage, taken by NASA astronauts.

As was covered in Chapter Two, there is a theory that the Moon is hollow and is an artificial satellite. If this theory is true, then we can assume that it was either built from the ground up, or it was a planetoid that was hollowed out, with the inside being transformed into a highly advanced, state-of-the-art ship! After its construction, for reasons that are obviously unknown, it was navigated across the universe, apparently on a course for Earth. Where it came from and how long it took to travel here is anyone's guess. However, when it arrived, it instantly changed the Earth. The Moon moving in would have had devastating effects on the Earth. It would have caused great chaos including flooding, earthquakes, and other catastrophic events. It has even been theorized that the Moon moving into place may have been the cause of the great deluge of biblical fame. Once the tumultuous activity ceased, the Earth settled into its position on its axis, became stable, and life on Earth began to thrive. When the Moon appeared, the seasons formed, tides were normalized, a new Earth was born! It became in some areas a paradise.

If this story is true, and there were extraterrestrials that constructed the Moon, then they were surely pleased with the results. We can only wonder about the intentions of those that may have brought the Moon here. Were they attempting to create a paradise on Earth? No matter how the Moon ended up here, no matter how it was created, it is important for us to learn more about it and perhaps try and connect the dots to its past and its connection to Earth. Given Earth's apparent vulnerability to catastrophic disasters, we as a species should endeavor to learn as much as we can about our past, as well as our potential place in this universe. If we are to survive as a species, then it is time for us to explore our position in the stars, and if we can, move forward.

It also appears that what we have been taught about our ancient history is not correct. Much of the reason that we may be in the dark about our history may be because of past calamities on Earth that have left us in a state of global amnesia.

Interestingly, in his works, the Greek philosopher Plato spoke of such a situation. In recounting the tale of Atlantis, Plato spoke of an ancestor of his named Solon. Solon was a respected statesman that lived in Greece, in approximately 630 BC. Solon enjoyed traveling. He relayed to friends that during a trip to Egypt, he had been told of a great ancient empire named Atlantis that existed well before the Egyptian civilization had come into existence. It was a priest of the Egyptian city of Saïs, in the western Nile Delta, named Sonchis of Saïs that gave Solon this information. He told Solon before the Egyptian civilization came to be, an ancient civilization (Atlantis) had been wiped from existence due to a series of cataclysmic natural disasters. The priest stated that it was 9,000 years prior to the time of him meeting with Solon! In *The Atlantis Myth* (1948), author H.S. Bellamy discussed Plato's story of Atlantis, explaining the events while incorporating Plato's discourse:

> When Solon visited Saïs he was received with every honour. Discussing problems of ancient history with certain priests who specialized in that subject he soon discovered that he, like any Greek, knew really nothing worth speaking of regarding matters of antiquity. On one occasion, desirous of getting the priests to talk about ancient times, Solon told them some of the oldest of our myths, the story of Phoroneus, the reputed 'First Man', and of Niobe, and the tale of how Deucalion and Pyrrha survived the deluge. He traced the line of their descendants and reckoning up the generations tried to compute how long ago these events might have happened. Thereupon one of the oldest priests spoke up saying; "Ah, Solon, Solon! You Greeks are like children. You know nothing of ancient things handed down by long tradition and you have no learning which is hoary with age. I will tell you why. There have been, and will again be hereafter, many and diverse destructions of

mankind. The greatest of these have been caused by water, while the lesser ones were due to many other causes. Take, for instance, the story, current also among you, of how once upon a time Phaethon, the son of Helios, attempted to drive his father's chariot, but was not able to keep it to the wonted track and so set fire to everything in the face of the earth, till he was annihilated by a thunderbolt. This sounds very much like a mere fable—but as a matter of fact it describes a deviation from their courses of the heavenly bodies which revolve round the earth, such as occurs after certain long intervals and causes destruction upon the earth through a great conflagration. At such occasions those who live high up in the mountains, in elevated dry regions, are more liable to extermination than those who dwell by rivers, or at the seashore...On the other hand, when the gods purge the earth with a deluge, the herdsmen and the shepherds in your mountains survive, while the inhabitants of the cities in your part of the world are swept away by the [waters...]. [There follows an obscurely worded passage the gist obviously is: Now, while you are subject to such dangers, we Egyptians are not only immune from the former calamity, living as we do in low-lying land, but we are also secure from the latter because the floods of our River Nile are always exactly predictable.] That is why the traditions preserved in our records are the most ancient which exist. ...Any noble, or great, or otherwise distinguished achievement that has come to pass either in your country, or in ours, or in any part of the world of which we have knowledge, has for ages past been recorded in the archives of our temples." If there is any truth to there being a time when mankind started over, then it is possible that we could have advanced enough to travel to the Moon, and lost all record or memory of it.

Ancient Alien Moon Theory

The ancient alien moon theory is the idea that extraterrestrials once had a civilization on the Moon, coupled with the thought that those same beings may have visited Earth. Ruins of their civilization on the Moon are believed to still be there, and a few are believed to have been photographed during NASA's lunar missions. There is also the speculation that the original lunar people, the "Selenites" are no longer there, and other extraterrestrials moved in. These alleged newer extraterrestrials could have been there for millions of years and it would still qualify them as ancient aliens. Some wonder how alien beings would have come to be on the Moon in the first place, and what their purpose is. A couple of theories have been put forward.

Theory 1) Extraterrestrials are native to the Moon and are simply living out their lives just as we do on Earth, with no special agenda. These beings may have been placed there by advanced beings that seeded the planets eons ago.

Theory 2) According to some ancient tales, the ruins of the civilization found on the moon were once a part of the Kingdom of Atlantis. Atlantis was believed to be an advanced civilization that existed on Earth but was destroyed. The people of Atlantis are thought to have had superior technologically, and had the ability to travel back and forth between the Moon and Earth in their spacecraft.

Ancient Tribes and the Ancient Aliens

There is an ancient legend found among the Zulu people of Africa that mirrors the Soviet scientists' hypothesis (as mentioned in Chapter Two) of advanced beings hollowing out the Moon and sending it on a trip across the universe to Earth. At the center of this story are two extraterrestrial brothers named Wowane and Mpanku who were known as the "water brothers." The brothers were the rulers of an ancient race of beings called the Chitauri. The Chitauri had scales covering their bodies that resembled those of fish. However, in lore, they are more commonly thought to have been an intelligent, humanoid reptilian species. According to legend, the brothers stole an egg (the Moon) from the "Great Fire Dragon," and removed the yolk rendering it hollow. They then proceeded to move the "egg" across the universe, eventually placing it in Earth's orbit. As a result of the Moon suddenly being so close in proximity to the Earth, there was great turmoil and chaos throughout the planet.

No one knows for certain where the story of the Zulu originated. What we do know is that there are ancient tribes of people that tell of being visited in the beginning of human history by advanced beings that they believed were gods. In modern times, these beings are thought to have been extraterrestrials. Could the Zulu people have obtained this fascinating account of the Moon's origin from visiting extraterrestrials? Is this the reason that their story so closely resembles that of the two Russian scientists' theory of the Moon's origin in today's modern era? Additionally, if we look at mankind's early history, we find tales of humans being visited by otherworldly beings and given knowledge on how to improve themselves and advance their life situation. They were educated in the fauna and the flora, agriculture, architecture, farming, sanitation and more. Some of these societies were given detailed information about the Solar System from these beings.

The Mayan people were given a map of the Solar System with information included in it that humans would verify centuries later. The Dogon people are a tribe whose legends offer an accurate location of celestial bodies that were not discovered by modern-day scientists until centuries afterward. In fact, the Dogon tribe believed from the beginning that their ancestors were descendants of extraterrestrials from the Sirius star system, located eight and half light years away. They were privy to extremely accurate, and very advanced information in the areas of astronomy and mathematics, given to them from the beings from Sirius. Researchers understand that this information was beyond the scope and knowledge of such a primitive people and are certain that the story of the Dogon being visited by otherworldly beings is an accurate one. It was clear that the Dogon could not have discovered this information on their own. Interestingly no one knows where the Dogon tribe originated. Some have surmised that they were of Egyptian descent, but this has not been proven. Remarkably, their rituals, observances and traditional ceremonies were aligned via the movements of an invisible and yet undiscovered star, now known as Sirius B. The Dogons had the knowledge that Sirius B journeyed around Sirius A every 50 years in an elliptical course. This information was not known by Western astronomers during that period. Sirius is a brilliant star sometimes referred to as the "Dog Star" or "Sirius

A." It is the most brilliant star in Earth's nighttime sky. In fact, the name Sirius is derived from the Greek language and means "glowing." Sirius shines so brilliantly that it was well known to ancient peoples. However, locating the companion star, now known as Sirius B, in 1862 astonished astronomers. Sirius B is 10,000 times dimmer than the brilliant Sirius A. However, the Dogon people had known of its existence for thousands of years. Interestingly, the Dogons were also aware that that Saturn has four moons and has rings around it. They had this knowledge even though these cannot be seen without powerful telescopes. The Dogons are believed to have been given this information via extraterrestrials known as the Nommo. Taking into consideration the tales of old that speak of beings from the stars coming to Earth, we might consider taking the story of the Zulu people into account when searching for the origin of the Moon. It just could be based on a real explanation of how the Moon came to be where it is.

Ancient Aliens Between the Moon and Earth

There are tales from our ancient past that speak of aircraft that could travel from Earth into space. One of the places that was visited was the Moon. In the past, people considered these stories to be fantasy and myth. More recently, due to increased interest in man's ancient origins, people are looking at the possibility that these stories may be true. If we examine the myths of the Native Americans, ancient European lore, the creation stories of the Australian native peoples, the Sanskrit texts of India, Sumerian lore, and others, you will see that all the major world cultures have one story in common, that is of humanoid beings arriving on Earth from the stars. These star people could obviously fly, and brandished powerful technology. In each tale, these extraterrestrials arrived in spacecraft. As a result, mankind had the knowledge of advanced aerodynamics in our prehistoric past.

There are many tales of civilizations that had spacecraft. Some could travel into space, to the Moon and beyond. What's startling is that these accounts can be found all over the world, with the majority and most detailed being found in Indian Sanskrit writings. In fact, records of ancient aircraft and spaceships, long before they were seen in our modern world can be found across several continents. There are many tales about the "gods" navigating these craft. There are references to spacecraft in the nearly five-thousand-year-old *Bhagavad Gita*, the *Ramayana*, and the *Mahabharata*.

In his article "Ancient Writings tell of UFO visit in 4,000 B.C.," John Burrows tells of flying machines in the Ramayana stating, "And in the *Ramayana* (writings), there is a description of Vimanas, or flying machines, that navigated at great heights with the aid of quicksilver and a great propulsive wind." *The Bhagavad Gita* tells of several gods from the cosmos traveling to the Earth in spaceships. There are also depictions in ancient artwork showing what appears to be saucers and other flying aircraft. This is evidence that people from that period viewed flying as a regular feature in their lives, even though they may not have experienced it, since it was the "gods" that were purportedly flying these craft. The most common name for these spacecraft is "Vimana." The word Vimana is Sanskrit. Over time it has held various meanings. In the present day, it means flying machine. Another meaning of Vimana is traversing.

The Vimanas performed at different levels. They came in a variety of shapes and sizes and were only piloted by "deities." Some were faster, and others could cover longer distances. Some were so advanced that they were capable of space travel and could reach the Moon and beyond. There were four categories of the Vimanas. These included the Conical Gold Ships, the Silver Rockets, the Three-tiered Airships, and the Sakuna crafts which were designed to resemble birds. There were one hundred and thirteen various types of these four classifications of Vimanas. For the most part, they were circular, saucer shaped craft. A passage

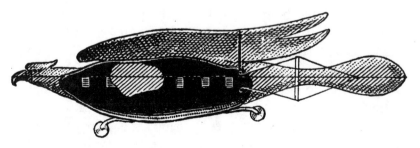

An artist's drawing of the bird-type Shakuna Vimana.

SHAKUNA VIMANA

HORIZONTAL SECTION

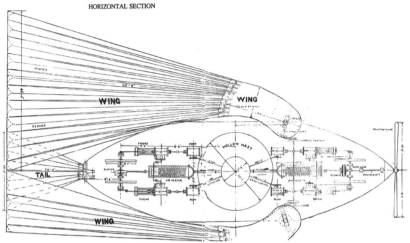

An artist's drawing of the bird-type Shakuna Vimana.

in the *Samarangana Sutradhara* states, "Strong and durable must the body of Vimana be made, like a great flying bird of light material. Inside one must put the mercury engine with its iron heating apparatus underneath. By means of power latent in the mercury which sets the driving whirlwind in motion, a man sitting inside may travel a great distance in the sky." Today we can only speculate if the "man" sitting inside was from an earthly advanced civilization, before a cataclysm destroyed it, or was it an extraterrestrial traveling back and forth between the Earth and the Moon.

The Hindu god Shiva is often portrayed in art as flying on his bird that was called Garuda. This depiction is believed to represent Shiva riding a spacecraft, perhaps that of a Sakuna. Garuda was known for flying to the Moon and for taking Shiva to various regions throughout the galaxy. The Hindu god Vishnu also rides on the back of Garuda. Some researchers studying the Indian tales of the Vimanas have concluded that they were very similar to what we consider a flying saucer. Interestingly, they deduce that they had stealth technology (the ability to make themselves invisible). Additionally, they are thought to have been adept at detecting enemy aircraft over long distances by using a type of psychometric radar. Needless to say, they were very advanced, far beyond what we have accomplished even today.

Although Indian lore carries the bulk of information on ancient flying machines, there are others that knew of aerial machinery, and may have been capable of space travel and quite possibly visiting the Moon. This is important because there have been constructions located on the lunar surface that are very similar to what we find in ancient Egypt such as the pyramids and sphinx. We should also consider that there could have been a sharing of information between some of the ancient cultures. Might there have been a sharing of technologies among the so-called gods? Could the beings in the ancient civilizations that were known as gods have been a part of a whole, separating themselves to establish the various civilizations, but all reporting back to one source, with that source being on or connected to the Moon? In Egypt there have been depictions of aircraft found in a temple carving in the city of Abydos. It was etched into a granite beam somewhere around 3000 B.C. (approximately 5,000 years ago). They are located on the walls of Seti I's temple. Within these carvings are images depicting what appear to be a modern-day helicopter and what has been referred to as a "futuristic airplane." They are known as the "Helicopter Hieroglyphs." There has been controversy over the authenticity of these carvings. However, there are researchers that believe these images are proof of flying vehicles in the ancient world. There are also 1500-year-old pre-Columbian relics that resembles modern-day airplanes. One is known as the "Golden Flyer." The pre-Columbians evidently understood aerodynamics, as a replica of the Golden Flyer was tested to see if it really had flying capabilities. The researchers were excited to find that the Golden Flyer could fly perfectly. One can only wonder what else the pre-Columbians has been exposed to, or what else they had seen.

In Chaldea, a copper chisel was excavated at Ur depicting what appears to be two modern-day space rockets complete with emissions emanating from the bottom, as if they were taking off. In Sumerian pictorial scripts, three separate but related objects are shown. When put together, they appear to form a three-part rocket ship. How did the people of ancient Sumer know about rockets? An interesting tale comes to us from China that also alludes to space travel in ancient times. This story lies with the discovery of what may be the first Moon rock on Earth. In 1725, one Father Duparc, a French explorer, found his way to the remnants of Hsiung-Nu, located in an isolated northern part of Tibet. During his discovery, he found ruins and several artifacts including monoliths, a tri-level pyramid, and an imperial palace decorated with depictions of the Sun and Moon. In the midst of all of this, Duparc noticed what appeared to be an odd, large, milky-white stone. He documented his discoveries.

Over two hundred years later, in 1952, during a Soviet scientific expedition to Tibet, Soviet scientists reviewed ancient writings that validated Duparc's original findings. The documents revealed that the once advanced city of Hsiung-Nu had succumbed to a fiery destruction. However, one of the documents held by the monks stated something startling. It said that the "milky-white stone" had been "brought from the Moon." Researchers question if this could have been the first Moon rock on Earth. Due to the high regard for honesty in the Tibetan culture in the ancient world, it is believed to be very unlikely that this would be a lie or a hoax. The document is therefore considered to be authentic and credible. Could this discovery be evidence that ancient men at one time travelled to the Moon? Some have even proposed that the accounts of the Indian Vimanas are accurate and that during some Earth-wide catastrophe, humans fled Earth and may have traveled to the Moon to escape. If we look at all the evidence given to us from the past, we can well imagine that if there were a civilization on Earth, and even more than one, that became advanced technologically, then traveled to the Moon, they could have set up colonies there. We can surmise that if there are beings on the Moon, there is a chance that they may be human. In other words, *could they be us?*

Star Wars

Somewhere in the far reaches of space, there is a possibility of a war going on… Scientists are claiming that there appears to be what looks like a war happening among the stars, as they cannot explain the anomalous, unnatural explosions they are seeing in deep space. An article titled "Nuclear Wars in Antiquity" (from the *Space2001* Website) states, "Five or six years ago an article in the *Herald Tribune* said there had been at least 80 unexplained explosions in deep space during the last decade alone! This had baffled many of the leading scientists and astronomers who were at a complete loss to explain the phenomenon!" The most sizable of these blasts happened "180,000 light years away outside of the Milky Way galaxy." The article quoted Nuclear Physicist and Ufologist Stanton T. Friedmann (1934-2019) as saying, "Tremendous activity of this sort could well be life out there involved in a war!" Could a nuclear war in space have happened in our Solar System eons ago and this is the reason we are seeing what appears to be ruins on the Moon that look as though they have been bombed? There are ancient astronaut theorists, lunar researchers, ufologists, and others who believe that the Moon may have been involved in a war in the past. This supposition comes from the shape that the ruins and artifacts were discovered in, and the amount of what appears to be artificial debris located there, as well as what may be vitrified glass. Some theorize that this war may have even been connected to Atlantis.

Others associate it with the Sumerian gods called the Anunnaki. Could there have been a war of the worlds somewhere in our remote past? If humans had the ability of spaceflight, could there have been a conflict at one time that led to the destroyed ruins that we see on Earth, the Moon and possibly Mars? If so, was the civilization on the Moon at that time destroyed? Were there advanced ancient civilizations on Earth that were destroyed

The "Ziggurat" in the crater Daedalus from Apollo frame AS11-38-5564.

Full version of Apollo frame AS11-38-5564 of the Dadalus crater with the white square indicating the loction of the Ziggurat-like structure.

in a cosmic war as well? How about Mars? Perhaps this was a war over the territories of the Earth, Moon and Mars? Might this have anything to do with the asteroid belt? Is this all connected to the evidence on Earth of a mysterious ancient atomic war? And finally, in the past, did we go too far on Earth with nuclear bombs and destroy ourselves, causing Earth civilizations to have to begin again? This may be the reasons that UFOs are so frequently seen around military installations on Earth. They may be watching our progress. There is a rumor that Apollo astronauts were "warned off the Moon." If that rumor turns out to be true, then could an ancient war have any connection to the idea that extraterrestrials do not want us on the Moon? It could be that Earth has a past with alien residents on the Moon that is not in our favor.

After creating the atomic bomb, Robert Oppenheimer (1904-1967) cryptically stated, "Now I am become death, the destroyer of worlds." This dreadful passage is from the Hindu writings of the *Bhagavad-Gita*. Oppenheimer certainly presents the correct

The "Ziggurat" in the crater Daedalus and a Ziggurat at Ur in Sumeria.

image of doom that an atomic bomb most certainly is. One wonders why we would find such a line in the *Bhagavad-Gita*. It is because there are tales in Hindu texts that speak of battles where weapons with extremely intense, high heat were used, such as the *Brahmashirsha astra*, a devastating weapon with the ability to obliterate life on Earth and possibly beyond. Researchers believe from these writings and other evidence that somewhere in mankind's past, some countries had nuclear arsenals.

If we step back into the warring darkness of Earth's past, we will find evidence of destruction from what appears to be nuclear weaponry long before it was reinvented in the modern era. Evidence of this can be found in South America where ruins of once magnificent cities can be found. These cities were discovered in the isolated and nearly impenetrable Amazonia, otherwise known as the Amazon Jungle. They were found during Spanish expeditions during their search for the "Seven Cities of Cibola." In their journals, the explorers from those travels recounted their discoveries and adventures. Some told of great metropolises that had been reduced to rubble from some past disaster. They noted that the structures seemed to have been melted by what appeared to have been incredibly high heat. There are researchers who believe that this is evidence that points to an ancient atomic weapon; melted stone buildings and vitrified glass are the result of it use. One account spoke of what was once a glorious city where no plant life grew. This was unusual since the city had been surrounded by a lush, overgrown jungle for centuries. Today we know that this can be the result of atomic radiation! Alternately, there were radiated corpses discovered in the Gobi Desert. This appears to be the result of atomic weaponry that was used on the citizens there. Where did humans obtain that kind of weaponry thousands of years ago? According to legend, there were advanced civilizations that had built vast, majestic cities and had nuclear bombs. We have clear evidence that ancient humans had nuclear capabilities. Are these vitrified ruins on both worlds symbols of a conflict between the two? Also, was Mars involved?

Since the creation of the atomic bomb in 1945, Earth has been inundated with UFOs. There are also those that claim to have been visited by extraterrestrials and given information to attempt to dissuade humans from using the bomb in an effort to save humanity and all living creatures here. It is even rumored that government heads were approached by extraterrestrials about the nuclear threat. Some of these extraterrestrials may be directly connected to our Moon because as you will see in Chapter Nine, it appears that the NASA Apollo missions were being observed by what is thought to be otherworldly beings. If there were an ancient war in space, life was lost. *Evidently, they are not going to let it happen again.*

One theory is that this war of the worlds was among ancient civilizations vying for power, and one of them may have been Atlantis. As most know, Atlantis was an ancient legendary, highly developed kingdom that was first brought to the world's attention by the Greek philosopher Plato. It was believed to have been an island continent located in the Atlantic Ocean, west of Gibraltar. There is some question as to whether Atlantis was real or a fable. From my own personal research, I do believe that Atlantis was a real place. Of course, the movies portray it as a mythological, magical place for children. However, history shows it to be a seafaring, powerful, advanced empire. Plato wrote that the island sank beneath the sea due to an earthquake. It is believed that the Atlanteans were able to fly and could travel to the Moon as well. Purportedly, they had the ability to fly using solar and electromagnetic energy and sent explorations to the Moon ages ago, eventually expanding their territory to the orb.

Therefore, some believe that the ruins found on the Moon may have been a part of the Atlantean empire. Among the remnants discovered on the Moon were what appeared to be vitrified and melted ruins of stone cities, thought to be due to a possible war fought there. Interestingly, similar ruins have been found on Earth in Death Valley, California and, as stated before, the Gobi Desert. Might there be a connection? According to legend, Atlantis had rulers for different parts of the empire. Could this have been a war due to a power struggle between the leaders? Could Atlanteans on the Moon have been making a play for independence?

Some have claimed that the officials in the 1960s arguing to send men to the Moon were interested in more than just a space race. It is alleged that some were already aware that there were beings dwelling on the Moon, and that one of the reasons for sending men to the Moon was to investigate this possibility. It has also been speculated that these beings had been there since ancient times and that they had influence over earlier civilizations on Earth. We have already established that there is scientific evidence of an ancient atomic war that took place on Earth, such as radiated ruins and skeletons. Reportedly, there are also nuclear bomb craters here on Earth that date back to ancient times. Carbon-dating and radiation tests have been done on bombed ruins and artifacts as well. Some believe that all of this is connected to what appear to be nuclear bomb craters on the Moon.

In fact, the first space probe to orbit the Moon (*Luna 10*) sent back an image of what appeared to be an ancient pyramid on the far side of the Moon, along with ruins such as walls and roads. Researchers have proposed that these may be traced back to the story of Atlantis having a territory on the Moon. As the tale goes, a war erupted on Earth, and eventually extended to space and to the Moon, and some say perhaps it may have even stretched to Mars. Evidence believed to support this story of a war was found by British scientists that came forward with their research indicating that several of the Moon's craters are fashioned precisely like atomic bomb craters found on Earth. In the article "U.F.O. and Reported Extraterrestrial on Moon and Mars," Steve Omar/En Mar writes:

> I have also talked to Farida Iskiovet of the United Nations U.F.O. investigation and 8 former army, navy or air force intelligence officers who had top secret security

clearances, as well as former N.A.S.A officials, and our department interviewed some ex C.I.A. agents... who all say they know there were ancient astronauts influencing earlier cultures... and here is the bombshell... there is scientific proof of ancient nuclear warfare on Earth left by radiated ruins and skeletons, nuclear bomb craters on Earth from ancient times, and buildings and objects with that nuked look, as well as carbon-dating and radiation tests on these things. Is that tied into the nuclear bomb appearing Moon Craters?

Bomb craters are unique in that they have a wall and floor construct that is completely dissimilar to craters created from meteors hitting the lunar surface.

It has also been suggested that the structures located near the Sea of Tranquility and dubbed the "Blair Cuspids" could also have a connection to the ancient war theory. William Blair, an archaeologist specialist for the Boeing Institute of Biotechnology, studied images received from the *Lunar Orbiter 2* space probe that were taken near the western rim of the Sea of Tranquility. The picture revealed a white obelisk that resembled the Washington Monument. It was estimated to be approximately 639 feet tall. Other obelisks were in the area as well. Could these have been remnants of the ancient civilization of Atlantis? *Some believe so...*

An Ancient Earth-Styled Moon City

In the *Star Trek The Original Series* episode titled "Who Cries for Adonais" (the second episode of season two) the god Apollo is alive and well and living in space in his own self-imposed seclusion. What is interesting here is the encounter with captain Kirk of the *Enterprise* and his crew. There Apollo reminds us that the gods once visited Earth, and that they were loved, admired, and worshipped, and they enjoyed it! Apollo wishes to go back to those old ways, but humanity has evolved. They no longer worship gods. This sad realization causes him to hurl himself into the wind and disappear forever. While it is true that mankind cannot go back to those old ways as we have evolved and we no longer recognize powerful entities as gods, we understand from this tale what could have happened in the beginning annals of our history. We were visited it is true. But these gods for all intents and purposes were spacefaring beings that somehow happened upon a primitive world (either purposely or unintentionally), and with either benevolent or malevolent intentions, changed our course of development as we paused for a time to worship them! Were they helpful to us? Yes. *But did they stagnate us as well...?* Perhaps.

It appears that these beings assisted in our development by educating us in how to thrive and live longer, healthy lives. They taught us about architecture, agriculture, the arts, cleanliness, and other productive ways of living. Their influence can be seen in the great civilizations of our ancient days such as Egypt, Rome, Greece, India, Mesopotamia, Sumer, and some religious traditions. If they had the means of space travel as it appears they did, then their influence may have reached to the Moon as well.

When we study the ruins and buildings located on the Moon there are some similarities to what we find in the ancient world. What we may well be looking at in the ruins located on the Moon is an ancient Earth-styled Moon city. If we look at the research which includes testimonies and photographs from NASA missions, we see there are what appear to be Grecian, Egyptian or Roman-styled huge naves, massively tall steles, cupolas, pyramids, and there is even a sphinx, all on the Moon. In other words, this old, ruined city on the Moon, appears to resemble those of ancient Earth. How is that possible? Is it a coincidence? Whether or not these beings on the Moon were the actual perpetrators of what almost appears to be a hoax on mankind, or if the Moon inhabitants were also in the same position as we Earthlings were when it came to these superior beings, we may never know. As far as we know these gods left Earth. Did they return to the Moon? Or did they leave the Solar System entirely?

Prophets of the Moon

Adam

Was the creation of the first man Adam, connected to the Moon? There is a tale from mankind's ancient past involving extraterrestrials that connects the Moon and Earth and the first created human. Some believe it to be a fable. Others hold that it is correct and from a time from which it has been forgotten. This account relates to the Sumerian gods known as the Anunnaki. According to Sumerian lore, the Anunnaki created the first human whom they called "Adamu." Adamu was also known as Adam, the first human created in Judeo-Christian-Islamic beliefs. The Sabaean culture asserted that Adamu originated on the Moon and was brought to Earth by his creators. Adamu, per the Sabaeans, revered the Moon and taught others to as well. This belief about Adam's association with the Moon has led to the idea that the Moon is really the planet Nibiru, the original home of the Anunnaki.

Due to this account, it is also thought that the Moon is a disguised spaceship from Nibiru, which is also known as "Planet X" and the "tenth planet." For this reason, some believe that Adam may have considered the Moon home, and the home of the Elohim (the gods). Moses Maimonides, an influential medieval Jewish philosopher once expressed the beliefs of the Sabaeans, writing, "They deem Adam to have been an individual born of male and female like any other human individuals, but they glorify him and say that he was a prophet, the envoy of the moon." Per the Sumerian tale, Adamu was created by two gods Enki and Ninki who were a part of the Anunnaki godship. According to Sumerian lore, the Anunnaki created a slave race to mine gold on Earth. That slave race was mankind. In the article titled "The Scariest Book of All Time: Cosmological Ice Ages a Short Explanation," Henry Kroll, the author of the book *Cosmological Ice Ages* writes, "The fact that Adam may have been preaching a worship of the Moon is just more evidence of his possible birthplace and the location of his acknowledged creators."

It is said that the Anunnaki in their mining for gold, stored it on the far side Moon. Once humans mined the gold on Earth and turned it over to the Anunnaki, the gold was transported onto cylindrical shaped spaceships, taken to the Moon, and stored there until it was transported to Nibiru. The reason behind the Anunnaki needing the gold was because their atmosphere was dying, and gold was lifesaving for their planet. If there is any truth to the Sumerian tale and the story concerning Adamu from the Sabaeans, then mankind may have a direct connection to the Moon by way of the creation of humans.

Could it be that mankind originated on the Moon? This of course is a stretch for most people. However, we must question what we think we know, and think outside of the box in order to find the truth. After a great rift between the gods over the fate of humans, the Anunnaki are believed to have left the Moon and returned to their home, the planet Nibiru. Today, some researchers believe that there are still Anunnaki on the Moon, residing in the interior of the Moon, all the while observing mankind's progress. Researcher and Author C.L. Turnage states in her book *ET's Are on the Moon & Mars* (page 7) that the Anunnaki, "were actually present during the time our astronauts made their historic Lunar landings." There is also a theory that the Moon is still being mined for gold. There is said to be missing gold on Earth. Nobody seems to know where the gold is disappearing to. It is believed by some that the gold is still being transported to the Moon and stored on the back side.

Enoch

One of the first prophets that we are told in ancient texts traveled to outer space is the biblical prophet Enoch. The Bible states that he was the son of Jared and the father of Methuselah. According to the account, Enoch lived an incredible 365 years before he was "taken" by God. Even though Enoch was an important prophet of the Bible, in our modern times some view him as an ancient astronaut due to his travels off Earth. There is an account of Enoch

An old print of the Biblical patriarch Enoch.

being taken to heaven by an "angel" in the *Pseudepigrapha*. There Enoch witnessed what he referred to as "angels" in classes learning about life on Earth. One questions whether these were angels or extraterrestrials. Some have speculated that this journey was really a trip to the Moon since it is in such close proximity to Earth. In another account, Enoch was carried to space by two large men that had "eyes like fire," who were described in the *Pseudepigrapha* as "the likes that had never been seen before on the Earth." These men, according to author C.L. Turnage in her book *ETs are on the Moon and Mars* (page 57), gave Enoch a view of the "power of the Moon's light." When referencing this account, Turnage takes it one step further asking, "Did Enoch have the unique experience of being the first earthling to see a Nibiruan's moon base?" In this account, it is possible that Enoch may have visited the Moon. One wonders if what we are reading in the *Pseudepigrapha* is really a description of extraterrestrials on the Moon learning about Earth. Some researchers believe that the word "angels" in ancient writings are really referencing extraterrestrials. It is believed that early Judaic beliefs had no other reference for powerful beings visiting Earth other than angels.

Enoch's travels also sound very much like those of the contactees of today that claim they were taken to space and visited other planets. The account of Enoch in the *Pseudepigrapha* tells of Enoch witnessing "angels" in classes studying the "flora and the fauna" of Earth. Could the beings that Enoch witnessed have been extraterrestrials learning about Earth from the Moon? Could they have been the, "Watchers" of old? Or perhaps they were Selenites (the original lunar inhabitants). Enoch's account just may give us a bird's eye view into an ancient scenario on the Moon, where the beings there were learning about Earth and its inhabitants. It certainly raises a lot of questions such as, Were they there to study Earth as a means of helping? Or are they simply studying for their own edification, just as we study the Solar System and planets today?

Even more interesting is the question of whether they were interested in an environmental rejuvenation on Earth. Were the "angels" that Enoch saw studying really extraterrestrials assigned to Earth to help improve life on the planet? If so, these very same beings that were watching and studying Earth from the Moon could have just come down, posing as gods and others, as a means of helping to forward humankind's development. There is one possibility that should be considered. That is the beings that took Enoch on his travels were intent on him returning to Earth to report what he had seen on the Moon. This may have been a means of alerting humanity to their presence. If that is the case, then the entire scenario seems to have been lost in translation in the centuries past... *until now*. In other words, we could have been alerted very early on that we are not alone in the universe, and there are those who are interested in us, and helping us, and that may have had a hand in the Earth becoming stabilized to support life here.

A Russian Zond 3 Lunar Probe photo of a huge tower on the Moon.

81

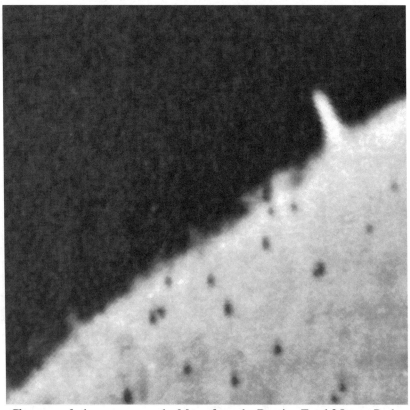

Close-up of a huge tower on the Moon from the Russian Zond 3 Lunar Probe.

Chapter Five

Strange Happenings

"Over the past centuries, literally hundreds of similar strange
lights and other weird happenings have been observed on this
purportedly dead world and reported."
—Don Wilson, *Our Mysterious Spaceship Moon*

There are many tales of strange events occurring in the vast,
cold emptiness of space. There is also no shortage of strange
happenings when it comes to the Moon. From those first years of
space exploration, researchers, scientists, astronauts, and others
have witnessed baffling, inexplicable phenomena on and around
the Moon, ranging from strange, unexplained lights and radio
signals to UFOs. It is believed by some researchers that someone
or something is behind these mysterious events. The question is,
"Who is causing all the commotion?" Many researchers agree that
the strange happenings are evidence that the Moon is inhabited
and proves that it is not just a large chunk of mass in the sky with
nothing going on. Could extraterrestrials have been visiting Earth
from the Moon all through human history? It is a possibility.
Especially if the Moon is a stopover point as proposed in Chapter
Three.

The Moon is an Augmented Satellite
It is a theory that has been recited time and again and is one
that is not going away. It is the theory that the Moon is likely an
augmented satellite. People think this is a ludicrous idea created
by conspiracy theorists. I talked about this topic a bit in Chapter
Two. It is appropriate to mention it here as well, to start us off on
the topic of the Moon's strangeness. In trying to understand the

incredible complexities of our Moon, two top Soviet Scientists, Mikhail Vasin and Alexander Shcherbakov of The Soviet Academy of Sciences, came forward with an astonishing theory: the Moon is a hollowed-out machine. Specifically, they believed it to be a spacecraft. They stated that they had examined the evidence, which included NASA's data, and had reached this conclusion. What could two top scientists who worked for the Soviet government gain by publishing such a report? They believed it to be true and put their reputations on the line by bringing it to the world's attention. The name of their paper was aptly titled "Is the Moon a Creation of Alien Intelligence?" They were not alone in their beliefs about this theory. In 1962 Dr. Gordon MacDonald, who was a NASA scientist, published a paper about the Moon. He is quoted as stating, "It would seem that the Moon is more like a hollow than a homogenous sphere." The prominent American cosmologist, Carl Sagan, once stated, "A natural satellite cannot be a hollow object."

Crater Strangeness

There have been a number of arguments to support the Moon being an augmented satellite. One example is the strangeness of the depth of lunar craters. The collision of meteorites on the surface of the Moon can release immense amounts of energy, some comparable with nuclear blasts. Here, a mystery arises upon examination of the Moon's craters. Even though they vary in diameter, they do not vary much in depth. No matter how large or how fast, the meteorites do not seem to penetrate the lunar surface more than 1.2 to 2 miles, even though they may leave behind a 40-mile-wide crater. Scientists are unable to explain this unusual occurrence. Professor Kirill Stanyukovich, a Soviet Physicist famous for his studies on lunar craters, was surprised at the results of his own investigations on the subject, stating, "A missile of a sizable character (say 6 miles in diameter) must, on collision with the Moon, penetrate to a depth equal to 4 or 5 times its own diameter (24-30 miles). The surprising thing is that however big the meteorites may have been which have fallen on the Moon (some have been more than 60 miles in diameter), and however fast they must have been traveling (in some cases the combined speed was as much as 38 miles per second), the craters they have left behind are for some odd reason all about the same depth, 1.2-2 miles, although they vary tremendously in diameter." The conclusion of the shallowness of the craters is that the Moon has an *impenetrable* hull. To some, this is an indication that the Moon was built in a way to withstand dangers such as meteorite bombardment. I wonder, if the Moon really was built, as some have proposed, then what is beneath the surface that is so precious to protect, that not even meteorite explosions can penetrate it?

Rocks

The Apollo missions, all total, brought back 2,196 rock samples extracted from the surface of the Moon. The weight equaled 842 pounds. Scientists have dated some lunar rocks to be 4.5 billion years old. This is puzzling because their age places them as 1 billion years older than the oldest rocks ever found on Earth. It also means that they are nearly as old as the Solar System. After analyzing some of the samples, scientists found them to have processed metals in their composition. The elements located included chromium, titanium, zirconium, brass, mica, neptunium 237, and uranium 236. These metals are not known to occur naturally, are mechanically strong and have anti-corrosive properties. Putting these together would offer strength and fortification against extreme temperatures, meteorite bombardment, cosmic radiation, and other potentially hazardous elements. Scientists are baffled by this, as these type of rocks and metals, they maintain, should not be on the Moon. Additionally, the rock samples taken from the Moon were also discovered to be magnetized. This was also problematic for scientists, and was yet another mystery, since the Moon has no magnetic field. Just where the magnetism came from is unknown. It does appear that whoever built the Moon (if indeed it were built), were incredible and amazing scientists. We are looking at what would be a highly sophisticated, powerful race of beings, far beyond our imagination. It is interesting that the scientists stop short of stating this, when the evidence is so revealing.

Orbit

Our Moon's orbit is extremely odd. It does not behave in ways that are typical for a natural satellite. The orbital distance of the Moon causes it to appear as if it is the exact size in the sky, as the Sun. Additionally, during a total solar eclipse, the Moon obscures the Sun almost exactly. The Moon is also in synchronous rotation. The *Free Dictionary by Farlex* defines a synchronous rotation as "The rotation of an orbiting body on its axis in the same amount of time as it takes to complete a full orbit, with the result that the same face is always turned toward the body it is orbiting. Earth's moon exhibits synchronous rotation." It turns on its axis in approximately the same time it takes to orbit the Earth. In addition, the Moon's orbit around the Earth is completely spherical. An interesting fact is that this unusual orbit would not have been captured by the Earth's gravity, as observed in the largely oval-shaped orbits of the other moons in our immediate Solar System. This very seriously implies that the Moon was brought here, and if it were, that means that we are not alone in the universe, and that we are dealing with something that is far ahead of us in technological achievements. It opens a whole new outlook as to our place in the universe. Perhaps the Moon was placed in this position, with so many abnormal characteristics, as a way to tell us that it was intelligently made. Therefore, the Moon itself a sign to us that we are not alone in the universe.

Lunar Eclipse

In centuries past, a lunar eclipse was considered a dark omen. If anything nefarious happened on the night of a lunar eclipse, the eclipse itself was blamed, as it was believed that the darkness was a curse. As a result, a lunar eclipse during those days was a source of great mystery for some, and a cause of fear for others. Today we know that a lunar eclipse happens when the Moon is obscured as it moves into Earth's shadow. During an eclipse, the Earth is just the right distance from the Moon to cast a large enough shadow to completely cover the Moon. Some find it strange that the distances between the Earth and the Moon, and between the Earth and the Sun, as well as the diameters of the Moon and the Earth, are relatively such that the Earth would block the Sun completely during a lunar eclipse. Some claim that this is too unusual to be a coincidence and maintain that it is evidence that the Moon is not a natural satellite. Due to the complexity of the eclipses, and the very small chance of everything lining up to create an eclipse, some lunar researchers believe that this is a sign for us from the creators of the Moon. Some maintain that they have been waiting for humans to advance enough to put the clues together, so that we would be able to discern that the Moon and perhaps the entire Solar System were not created by chance, but by design from advanced extraterrestrials. There is a theory that the originators are waiting for us to put this all together, and understand that we are not alone, and that our beginnings started with beings from a very advanced world.

Strange Lights

Interestingly, there are accounts from astronomers that mention witnessing strange lights on the Moon during a lunar eclipse. No one knows the source of the lights, only that their appearance is mysterious and beyond explanation. Frederick William Herschel (1738–1822) reported seeing as many as 150 lights in various areas of the Moon during a lunar eclipse. In his book *Flying Saucers on the Attack* (page 217), John Wilkins writes about the observation by Herschel stating, "The famous astronomer Frederick William Herschel, was looking through a 20-foot reflector telescope, on October 2, 1790, when he saw, in time of a total eclipse of the moon, many bright and luminous points, small and round."

German American Astronomer Joseph Zentmayer (1826–1888) was perplexed to see objects on the Moon during a lunar eclipse. The objects were illuminated and moved in parallel across the Moon. In the early 1800s, Astronomer John Herschel also observed abnormal lights that were travelling across the Moon during a lunar eclipse. The "Chronological Catalogue of Reported Lunar Event" is one of the primary sources of this information, documenting several accounts of mysterious lights on the Moon during a lunar eclipse. In 1847, there were reports that bright lights had been seen on the Moon during a lunar eclipse. (The Aristarchus crater is well known for mysterious activity.) On both December 5, 1881 and December 27, 1898, each time during a lunar eclipse, astronomers reported that the Aristarchus crater was completely lit up. The same occurred on May 23, 1891, with the exception being that the crater was reportedly illuminated towards the end of the lunar eclipse.

On February 19, 1905, during a lunar eclipse, a bright spot of light was seen in Aristarchus. It was described as resembling a "star." Again, on August 4, 1906, October 7, 1949, and June 25, 1964, during a lunar eclipse, astronomers reported that the entire crater was illuminated. The Grimaldi crater has also been named as a place on the Moon that has had unexplained light activity. One of these incidents occurred during a lunar eclipse on June 25, 1964, when a brilliant flash of light was seen towards the edge of the crater. As far back as December 10, 1685, an odd red streak of light was seen on the floor of the Plato crater during a lunar eclipse.

There have been reports of lights during lunar eclipses in the Tycho crater as well. In two of the cases, the entire crater was illuminated. The dates were August 15th, 1905 and April 1, 1901. On November 7, 1919, a glowing light was observed in the vicinity of the crater throughout a lunar eclipse. Amazingly, in 1912, astronomers in both France and Britain that were observing a lunar eclipse saw what one described as a "superb rocket," burst forward from the lunar surface. To date, there have been no explanations of what the strange lights seen during the lunar eclipses were. Also, what could the objects crossing the Moon have been? We can only wonder how many of these occurrences happen today that we are not informed about.

Where there is Light, there is Life!

One of the strangest phenomena associated with the Moon is the mysterious and perplexing lights that have been seen on the Moon for centuries. Hundreds of these illuminations have been reported from reputable astronomers, scientists, astronauts, and Moon observers all over the world. No one knows where they come from nor the cause of them. They are famously known as *Transient Lunar Phenomena* (TLP) and are also referred to as *Lunar Transient Phenomena* (LTP). Transient Lunar Phenomena is a term coined by British astronomer Sir Patrick Moore in the 1960s. Moore witnessed strange lights that he could not explain. One that especially baffled him was located in the Gassendi

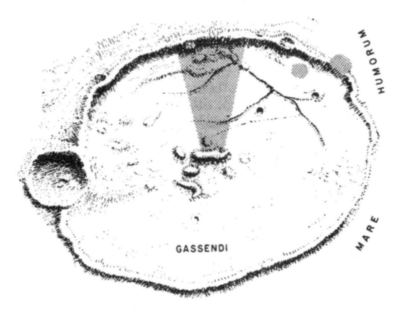

A drawing of the light phenomena seen at the Gassendi crater.

crater. He stated that it was of a reddish color and appeared to be glowing. At that time, he described what he had seen as "a color phenomenon." He observed it from April 30–May 1, 1966. Moore was not the only one that has been bewildered by these lights. In fact, ever since the invention of the telescope, astronomers have witnessed TLPs on the Moon. Many have recorded this activity. These lights are one of the reasons that early scientists believed that the Moon is inhabited.

The lights are random. They appear in various areas across the lunar surface, with no set time or pattern. TLPs are seen on the Moon so often that in 1966, NASA commissioned a catalogue to be compiled of anomalous lights on the Moon, with detailed information that identified where the lights were located, the time, description, date, and the name of the observer. The report is titled "Chronological Catalogue of Reported Lunar Events (also known as "NASA Technical Report TR-R277"). The catalogue was composed and written by four of the top scientists and astronomers of that time. They included Barbara Middlehurst of the University of Arizona; Jaylee M. Burley of the Goddard Space Flight Center; Patrick Moore of the Armagh Planetarium; and

Barbara L. Welther of the Smithsonian Astrophysical Observatory. The catalogue references four centuries of sightings recorded by over 300 astronomers. It consists of 570 lunar anomalies, including strange lights, unexplained lunar events, and mysterious phenomena observed on the Moon.

It should be noted that NASA chose these 570 events from over 2,600 such events that had been observed. The catalog chronologically lists Moon abnormalities that were reported by astronomers from 1540 to 1967. According to the report, "The purpose of this catalog is to provide a listing of historical and modern records that may be useful in investigations of possible *activity* on the moon." The report explains that TLPs are irregular, short-lived flashes of light and unexplained illuminations on the Moon. Specifically, the report is referring to all anomalous lights, transitory changes, unusual colors on the lunar floor, glowing mists, flashes of light, and craters with what appear to be lighted domes. There was even one instance when lightening was seen, prompting some to speculate about there being an atmosphere there.

The report was released to the public in 1968 and is still available today. What is fascinating is the word "activity" as used in the report. Activity on the Moon is much denied by NASA today. Somewhere along the line, they had a change of heart about reporting strange happenings going on with the Moon to the public. In fact, the very idea that they kept track of these events and then later commissioned this catalogue shows curiosity and perhaps even concern about these odd lunar events. And why wouldn't they be concerned, considering they were planning to send people to the Moon?

These TLPs have been witnessed in different areas of the Moon. They last for various periods of time ranging from a few minutes to several hours. We know that they are not due to volcanic activity as there has been no volcanic activity on the Moon for three billion years. They come in too wide a range of forms to be electricity. So, what other explanation is left? Might they be intelligently made? Is it possible that NASA commissioned the *Chronological Catalogue of Reported Lunar Events,* because they were alarmed that there just may be someone on the Moon that is causing them? Eventually, they made the incredible decision to send men to the Moon, despite the TLPs.

Although TLPs had been seen for centuries, it is the experience of cartographers Edward Barr and James Greenacre, of the United States Air Force Aeronautical Chart and Information Center (ACIC), who unintentionally inspired today's interest in the phenomena. On the evening of October 29, 1963, the two were working at Lowell Observatory in Flagstaff, Arizona. Their job was to map the Aristarchus crater, which is approximately twenty-seven miles in diameter.

Both Barr and Greenacre had studied the crater and knew that entire area of the Moon very well. That evening, the two were amazed to see what they described as intensely glowing, reddish-pink lights on the lunar surface, near the southwest interior rim of the Aristarchus crater.

In an article for *Time* magazine titled "Astronomy: Spots on the Moon" (Friday, December 27, 1963), Greenacre stated, "I had the impression that I was looking into a large, polished gem ruby."

The sighting lasted less than twenty minutes. After they reported what they had seen to Dr. John Hall, the observatory's director at the time, he reported the sighting to "astronomical authorities." He also ordered "a close watch be kept on Aristarchus." Hall and other lunar observers were apparently intrigued by the sighting in Aristarchus as well. According to *Time* magazine, "in late November, two days after the edge of sunlight reached Aristarchus again, Dr. Hall and four other observers saw a reddish area, twelve miles long and 1½ miles wide, inside the rim right where one of the spots had been seen in October."

The sighting of the spectacular light anomaly as seen by Greenacre and Barr, who were well respected and considered at the top in their profession, garnered the attention of others in the field, and inadvertently inspired greater interest in the observance and study of transient lunar phenomena. Later, a report titled, "Lunar Color Phenomena, ACIC Technical Paper No. 12" (issued by the United States Air Force Aeronautical Chart and Information Center) went into detail about what Barr and Greenacre had witnessed on that evening in October of 1963. The department felt that the sighting was significant and of great importance.

However, although there were speculations, no real conclusion was ever made as to the cause of this exciting light phenomenon. In 1965, NASA started an investigation designated Operation Moon Blink for the exploration of unusual phenomena on the lunar surface. Here NASA requested observatories from all over the world to monitor and photograph the Moon. Within a few months of *Operation Moon Blink* being implemented, twenty-eight TLPs were identified, reported, and documented.

Additionally, former NASA astronomer Winifred Satwell Cameron created the largest database of TLPs ever published. She compiled 900 accounts of TLP sightings ranging from 1540 to 1970. Her collection included kinds of mysterious lights including glowing lights, strange shadows, flashes of light, lights moving across the surface of the Moon and more. Her work is still being used today in research and discussions about TLPs. Over sixty years after the Greenacre/Barr sighting, the question of the sources of these strange light occurrences remains unanswered, *at least officially.*

Early Sightings of Lights

One can only imagine what went on in the minds of those early astronomers observing odd lights on the Moon. They clearly knew they were out of place. They may have associated them with Moon inhabitants. Electricity had not been discovered yet; the light bulb wasn't used until the 1800s. Airplanes were not flown until the early 1900s, therefore one can only wonder what these early astronomers thought about the light phenomena, as they attempted to figure out what their origin was. One early reported sighting occurred in 1587, on a night of a crescent moon. A crescent Moon is the image of the Moon as it is seen in its first quarter, when it is in a bowl shape with edges ending in points. These points are sometimes referred to as "horns of the Moon." During this period, it is said that the Moon appears to be "smiling." In 1587, an astronomer in England reported seeing a peculiar light shining brilliantly inside the crescent Moon. He explained that it was in the area between the "horns of the crescent Moon." The light was highly unusual, and the astronomer knew that it should not have been there, as it had not been observed previously.

My first discovery of an early reported TLP that I found on this very strange journey down the rabbit's hole, is that of Minister Cotton Mather (1663–1728). In November of 1668, Mather observed an atypical light on the Moon. The light was strangely out of place. Stumped as to what it could be, Mather reported the sighting to the Royal Society in Boston in a letter dated November 24, 1712. Mather wrote, "ye star below ye body of ye Moon, and within the Horns of it... seen in New England in the Month of November 1668."

Mather knew the light was out of place. British astronomer Nevil Maskelyne (1732–1811) is known for being the first to scientifically gauge the Earth's weight.

In 1794, Maskelyne witnessed a mysterious cluster of lights traveling across the dark half of the Moon. In his case, he witnessed several lights traveling together. Whatever these lights were, Maskelyne knew they were out of place and strange. One can only wonder whether the people of these early days had any ideas about space travel, UFOs, and extraterrestrials. Another early reporting of an odd light was that of German astronomer Johann Schroeter (1745–1816) on September 26, 1788. Schroeter, who was the founder of selenography and who dedicated his life to the study of the Moon, reported to Johann Bode's *Astronomisches Jahrbuch (Astronomic Yearbook)* that he had witnessed a brilliant white point of light on the Moon near the bottom of Mont Blanc. He reported that it had a radiance of a 5th magnitude star, and that it had lasted for a startling fifteen minutes. This was no speck of light that suddenly appeared and quickly disappeared. Schroeter was startled. He could not figure out what he was looking at and reported the anomaly.

On the evening of March 7, 1794, William Wilkins witnessed a radiant light near a darkened area of the Moon. He is quoted in Harold T. Wilkins' book *Flying Saucers on the Attack* (1954), describing what he witnessed that evening stating, "This light spot was far distant from the enlightened part of the moon and could be seen with the naked eye. It lasted for 15 minutes and was a fixed and steady light which brightened. It was brighter than any light part of the moon, and the moment before it disappeared, the brightness increased." What could this light that appeared, shined for a quarter of an hour, and increased in intensity have been? Could it have been an extraterrestrial spacecraft?

The renowned German-British astronomer Frederick William Herschel (1738–1822), who has been hailed as the "greatest Moon observer in the history of the science," reported seeing several unexplained lights on and around the Moon during his career. In one case, Herschel reported seeing as many as one hundred and fifty lights in different parts of the Moon while observing a lunar eclipse. In his book *Flying Saucers on the Attack* (page 217), John Wilkins writes about the observation by Herschel stating, "The famous astronomer Frederick William Herschel, was looking through a 20-foot reflector telescope, on October 2, 1790, when he saw, in a time of total eclipse of the moon, many bright and luminous points, small and round." Herschel also stated that he had seen several lights traversing the Moon through the years. In November of 1821, he reported seeing irregular lights three times. Could these lights that were seen in the Moon's sky during eclipses and in the presence of the crescent Moon, have been extraterrestrial ships? We know today that there are UFOs around the Moon. Men of those early periods suspected that the Moon was inhabited. It is not clear that they imagined flying saucers, as spaceflight was still unknown to them; they must have been truly baffled, just as we are today.

On July 15, 1888, the then director of California's Lick Observatory, one Professor Holdon, reported seeing an unusual light. According to Don Wilson in his book *Secrets of our Spaceship Moon*, Holdon described it as an "extraordinarily and incredible bright [light] … the brightest object I have ever seen in the sky." Wilson goes on to state that Holdon found the extreme brilliance of the light hard to comprehend and thought that it may have been the result of volcanic activity. However, as noted previously, the Moon has not had volcanic activity for billions of years. German American astronomer Joseph Zentmayer (1826–1888) renowned for creating microscopes and other ocular apparatuses, observed baffling objects on the Moon that he could not account for. The objects were intensely bright and traveling in parallel across the Moon throughout a lunar eclipse.

Bright Spots of Light

A mysterious category of lights that still falls under the TLP category are those referred to simply as, "bright spots." The term is used often in the description of these type of lights seen around the Moon. These bright spots are small, illuminated spheres that have appeared on the lunar surface as well as in the skies above the Moon. They have been seen and documented by astronomers for centuries. What these spots are remains a mystery but they are prevalent enough to have sparked a conversation between the astronauts of the *Apollo 17* mission on more than one occasion. In one instance, astronaut Ronald Evans, the command module pilot states, "Hey, I can see a bright spot down on the landing site where they might have blown off some of that halo stuff." While working near the Grimaldi Crater (an area where mysterious light activity has frequently been spotted), again during the *Apollo 17* mission, astronaut Harrison Schmitt observed a bright spot of light. Excitedly Schmitt exclaimed, "I just saw a flash on the lunar surface!! Location: north of Grimaldi. It was a bright little flash near the crater at the north edge of Grimaldi." There are ufologists and lunar researchers who believe that these lights are the result of extraterrestrials, and the conversations between the astronauts prove that extraterrestrials were present as the astronauts went about their work on that final journey to the Moon.

Rays of Lights

Unexplained rays of light have often been seen radiating from craters on the Moon. They have bewildered scientists for years. No one knows what they are, nor what is causing them. They are regarded as one of the Moon's greatest mysteries. The rays vary in length and move in different directions. Lunar scientists speculate that the rays may be made up of dust-like substances. Some are approximated to be more than 1,500 miles long, with a majority measured at approximately ten miles wide. They have been viewed emanating from craters, including Aristarchus, Kepler, Copernicus, and Tycho, as well as others.

In April 1934, *The Marine Observer* retold a story that was published in the *Reader's Digest* book *Mysteries of the Unexplained* (page 254). The write-up reports that travelers aboard the ship S.S.

Transylvania, were in the North Atlantic at the time of the sighting. Moon observers that night had already seen the incredible auroras some hours prior to this strange event. The passengers aboard the ship that night witnessed what they described as an "orange ray" shooting upward from the Moon. The odd effect continued for fifteen minutes. This occurred on the night of May 2, 1933. It left witnesses perplexed as to what they had seen.

Transient lunar phenomena to date defy explanation. Who or what is causing these unusual events? Do they originate by natural means? If so, why can't the scientists give a plausible explanation? Is it because someone is responsible for them? Could author George Leonard have been right when he proclaimed that "Somebody Else is on the Moon?" Should we consider this theory viable, that there just may be someone up there responsible for the many years of strange occurrences on the Moon? There seems to be no other explanation for lights moving in various locations across the orb. In his book *Ancient Aliens on the Moon* author and researcher Mike Bara states, "The longest lasting light show ever recorded continued steadily for two days, *possibly a case of forgetting to switch off!"*

Strange Clouds, Fog and Mists

Clouds, fog, and mists have been seen on the Moon by astronomers around the world for decades, sparking some researchers to believe that the Moon may have more of an atmosphere than we know (or have been told). There are others that speculate that beings on the Moon may be purposely concealing what is there by means of spraying what appears to be something akin to a fog machine. One astronomer claims to have observed UFOs traveling across the Moon spraying a type of substance that he believes may be hiding structures on the Moon. He is not the first to have stated this hypotheses. However, those are only hypothesis that fall under the conspiracy theory umbrella. What we are seeing and what I am calling them could be totally different animals. They could be a form of gas emanating from beneath the surface. This could especially be the case if we really are looking at an inhabited Moon; some researchers maintain that there may be an energy source beneath the lunar surface that is being used by someone and may be seeping up and out to the lunar surface.

It is obvious today that long before humans began seriously studying the Moon (as opposed to worshipping it) there were strange events going on there. One of the first recorded sightings of a cloud formation was seen by the prominent Italian astronomer Gian Domenico Cassini (1625–1712). Cassini is connected to several astronomical discoveries and developments, such as the first study and analysis of Saturn's moons. The Moon's Cassini crater is named in his honor, as well as the *Cassini* spacecraft. In 1671 Cassini witnessed an unusual event on the Moon that remains a mystery to this day. As he raised his telescope that night, he was puzzled to see what appeared to be a white cloud-like formation hovering over the lunar floor. Although he offered no explanation as to what it may have been, he made certain to document the anomaly. Equally strange was the report from an unnamed astronomer in 1826 that witnessed what appeared to be a huge *black* cloud moving across the Sea of Crises (Mare Crisium). Years later, there was another reported sighting of what was described as a bright cloud on May 15, 1864 in the same area. The Plato crater, known as a "hotspot" for Moon anomalies for its high incidence of mysterious lights and other unexplained occurrences, is noted in NASA's "Chronological Catalogue of Reported Lunar Events" as a place where astronomers have seen cloud formations on several occasions. One was observed as far back as April 10, 1873. On that day the sun was high and two dim clouds were seen in the crater's area to the west. On October 5, 1878, a thin, shimmering light-colored cloud was observed. On April 15, 1932, a light-colored spot that resembled a cloud was observed as well. What in the world was causing it? If these are natural occurrences, why is it so difficult for scientists to explain the cause of these events?

Mysterious activity has also been witnessed in Schroeter's Valley. Schroeter's Valley (also Schroter's Valley; Vallis Schroteri) is a sinuous rille located on the near side of the Moon. It is named after the notable German astronome Johann Schroeter. On September 16, 1891, American astronomer William Henry Pickering spotted something unusual in the valley. In NASA's "Chronological Catalog of Reported Lunar Events," Pickering's

The Aristarchus crater with the Herodotus crater on the right and
Schroeter's Valley, a rille coming from Herodotus. (NASA)

description of what he saw that day is documented. He writes,
"Dense clouds of white vapour were apparently arising from its
bottom and pouring over its SE [IAU: SW] wall in the direction
of Herodotus."

What could have caused what appears to be an outpouring of
a thick white vaporous cloud to emanate from the walled area of
a crater? On February 10, 1949, an astronomer by the name of
Thornton observed what he described as a "Diffuse patch of thin
smoke or vapor from W side of Schroeter's Valley near Cobra
head, spreading into plain; detail indistinct, hazy (surrounding area
clear)." So strange! The Alphonsus crater is an impact crater that
dates back to the pre-Nectarian period. It is a whopping seventy-
three miles in diameter and can be located on the eastern edge of
the Sea of Clouds (Mare Nubium).

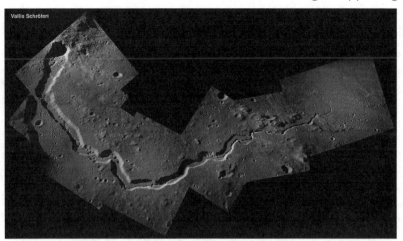

Schroeter's Valley, a rille near Aristarchus and Herodotus craters. (NASA)

On November 19, 1958, an astronomer watched as an odd cloud-like arrangement floated above the crater's central mountain area for a half hour. The prominent Russian astronomer Nikolai Alexandrovich Kozyrev (1908–1983), who was an astrophysicist with the Pulkovo Observatory and the Crimean Astrophysical Observatory, made the observation of the luminous "cloud." At that time, it was thought by some to have been caused by volcanic activity in the area. We now know that it is not possible that it was caused by a volcano, so then, what produced it? Could there be an energy source in the area? Could it come from some sort of industry, emitted from beings that inhabit the Moon? The Littrow crater is an impact crater named after Austrian astronomer Joseph Johann von Littrow. On September 13, 1959, astronomers were eager to photograph the Littrow crater and the area around it. However, they were unable to carry out their task due to an enormous blackened, cloud-like mass that was obfuscating the crater!

The Aristarchus crater has had considerable strange activity surrounding it, including baffling clouds materializing. According to sources there were two separate occasions, well over a century apart, where what appeared to be either mist or fog covered the crater. Strangely, a violet-colored fog has been reported around Aristarchus a few times. It is beyond imaging what it could be produced from. On August 6 and 7, 1881 what was reported as an intense violet light covered the region, giving the appearance of

99

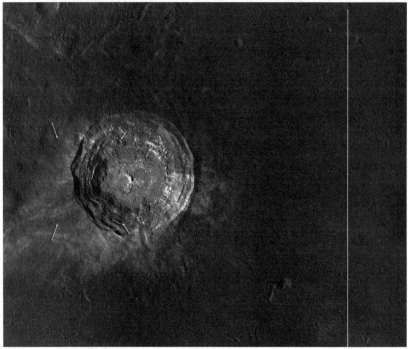

Aristarchus crater seen from directly above. (NASA)

the area being covered with fog. On October 2, 1955, Aristarchus appeared obscured in some areas due to another violet-colored haze. Although there are other reports of fog and haze and even mists being seen on and around the Moon, this is the only one that I know of involving such a strange color found in what appears thus far to be one location. It appears that something extraordinary may be going on inside of that crater.

The Grimaldi crater too has had quite a bit of anomalous activity over the years. Strange lights were covered earlier, but here we also have the one and only case of an odd "mist" reportedly seen there. This occurred on June 24, 1839. Astronomer Franz Gruithuisen observed what he described as a "grayish-colored" mist near the Grimaldi crater.

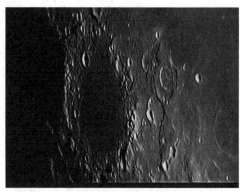

The Grimaldi crater.

Grimaldi crater and other surrounding craters.

The Sea of Tranquility (Mare Tranquillitatis) is a lunar mare located in the Moon's Tranquillitatis Basin. What makes this area of special importance is that it was the landing site of *Apollo 11,* the first mission to place men on the Moon, on July 20, 1969. There are several mysterious stories and events associated with the Sea of Tranquility. Odd constructions in the shape of obelisks dubbed the Blair Cuspids are located there. Also, there are the strange tales of the *Apollo 11* astronauts seeing otherworldly spacecraft there, and a theory that NASA sent the astronauts to the Sea of Tranquility specifically because they were aware of anomalous activity there. It is no surprise that there are reports of strange cloud-like materializations in the area. One source tells of a strange dark cloud-like mass being observed there. Could it be possible that this strange formation has something to do with the purported events that happened during the *Apollo 11* mission? On September 11, 1967, Canadian astronomers witnessed a dark cloud with violet edges moving slowly across the Sea of Tranquility. Is it possible that this eerie dark cloud with the violet-colored edges was extraterrestrial related? Soviet scientists Mikhail Vasin and Alexander Shcherbakov explained that their theory that the Moon

101

is a spaceship was one way of answering many of the perplexing questions associated with the Moon. Could this be the case with the clouds, fog, and mists scenarios? Are they associated with the theory that the Moon is an artificial satellite and possibly a sophisticated piece of machinery?

Unidentified Flying Objects (UFOs)

There are many cases past and present where UFOs have been spotted in various areas on the Moon. It is so busy up there that we need no longer even question the reality of them. There are now plenty of photographs and videotaped evidence for us to know that something or someone is up there, and what we are referring to as a UFO is in most cases, a spacecraft. In fact, several of the Apollo missions had encounters with UFOs. These encounters were documented and are well covered in Chapter Nine of this book. UFOs on and around the Moon have been seen flying, hovering, spraying (an unknown substance) and entering and exiting what appear to be large crevices on the Moon. These UFOs have been seen in various shapes and sizes around the Moon, much like the ones observed on Earth, which leads me to speculate that the ones we see here are coming from the Moon.

There are theories as to who is flying these craft and where they are from. These theories include:

•Lunar inhabitants
•Beings from several other worlds visiting the Moon
•Humans on a clandestine mission from Earth currently dwelling on the Moon
•Extraterrestrials on the Moon and Mars that are collaborating

It is believed that they may come from a variety of worlds. Some theorize that the UFOs are craft that are stopping on the Moon first, before visiting Earth, Mars and other cosmic objects in the Solar System including asteroids. It is also believed that these craft may be reaching the Moon via space portals, stargates wormholes or some other technology unknown to us that allows them to achieve shortcuts through space. These of course are things we know little about on Earth and have only seen in our science

George Adamski with Long John Nebel who holds a photo of a mothership.

fiction. Some researchers have even stated that they believe there are spaceports on the Moon. There are thought to be meetings going on there and in the same line of thought, it is believed that ships are coming in from other areas of the galaxy using the Moon as a kind of stopover point (as mentioned in Chapter Two) or even perhaps a type of visitation center.

George Adamski (1891–1965) was a very famous and sometimes controversial extraterrestrial contactee who claims to have visited the Moon via an extraterrestrial that had contacted him. Adamski claimed that this extraterrestrial was of Nordic origin. The account of what he allegedly witnessed on the Moon is eye-opening, if true. Among other things, Adamski alluded to the idea that extraterrestrials were regularly stopping through and visiting the Moon for all sorts of reasons.

Adamski, a Polish born American citizen, was one of the world's most famous extraterrestrial contactees. He served in the army before and during World War I. He was a philosopher, author, extraterrestrial researcher-investigator, photographer, and astronomer. During the 1950s Adamski became known for his interest in spaceships and spent a great deal of time searching for them. He documented many of his sightings of spaceships and validated his findings by means of photographs and video footage. He also had witnesses vouch for him, which lent credibility to his sightings.

Adamski was very much interested in the Moon, as he claimed to have witnessed several UFOs flying in the direction of the Moon during his search for spaceships. In an article published in *Fate* magazine in 1951, Adamski spoke of seeing spaceships that appeared to land on the nearside of the Moon. He noticed that others flew towards the far side and would vanish. In the article, Adamski stated, "I figure it is logical to believe that spaceships might be using our moon for a base in their interplanetary travels." After being contacted by extraterrestrials, Adamski was given the opportunity to travel to outer space. On two of those trips, they orbited the Moon. Adamski was given information about the Moon from the extraterrestrials. In his writings, Adamski shared what he was told and what he had witnessed when looking through telescopic equipment on the spacecraft.

Adamski claims to have observed enormous air docks on crater floors that were designed to hold large spacecraft. He was informed that visitors to the Moon had to first depressurize, for their bodies to become adjusted to the lunar environment. On the far side, Adamski described seeing hangars shaped like domes for spacecraft coming in that were carrying supplies. Apparently, there was a barter system in place where supplies were traded for minerals from the Moon. This indicates that the Moon is also acting as a kind of trading post, among other things.

With all the talk today about UFOs, USOs, UAPs, presidents being visited by extraterrestrials and more, we can certainly entertain the possibility that there are those that may have had extraterrestrial contact and be aware of things that we are not privy to. Regarding the many UFOs spotted around the Moon, the most common theory is that these ships belong to beings that are residing on the Moon.

However, could it be that non-lunar beings are sometimes briefed by lunar inhabitants on the Moon

An image of a cylindar-shaped object on the Moon taken by one of the Lunar Orbiters.

before arriving on Earth? Interestingly, the navigators of these spacecraft are not secretive when it comes to showing their ships and displaying their superior technology. They are however, shy about showing themselves. Could extraterrestrials be ready to introduce themselves, and that is the reason for the enormous increase in UFO sightings around the world today? I wonder if they are watching for the day of first contact, just as we are. *Do they have operatives here waiting to inform them of a good time to approach?*

In an article written by Steve Omar titled "U.F.O. And Reported Extraterrestrial On Moon and Mars," he tells of a city-ship being seen over the Moon. Omar writes:

> During the 1950s many U.F.O.S seen over Earth were tracked back to the Moon by government tracking stations in secret complexes in deserts in Arizona and Nevada and inside underground mountain bases. WE HAVE ONE PHOTO OF A SAUCER SHAPED CRAFT HOVERING OVER THE MOON, taken by a civilian astronomer. Sergeant Willard Wannail, who investigated U.F.O. landings in Ohio when in Army Intelligence, showed us an 8 by 10 CLEAR glossy detailed photo of a silvery spaceship hovering directly over a huge Moon landscape, estimated to be several miles long, and said to be CITY-SHIP designed to transport thousands of people between solar systems or galaxies and live for extended periods of time in self-sufficient orbiting communities!

What I find amazing about this information is that the term city-ship was used by the sergeant. One rarely hears the term city-ship, as it is considered to be a creation of science fiction. There has been mention of this type of ship from information that people have claimed were given to them from advanced extraterrestrials. To hear of someone in the Army using the term is simply astounding. It shows more may be known out there about extraterrestrials than the public has been told. This information also tells us that the Moon just may be a stopover point after all if there is a city-ship in the vicinity.

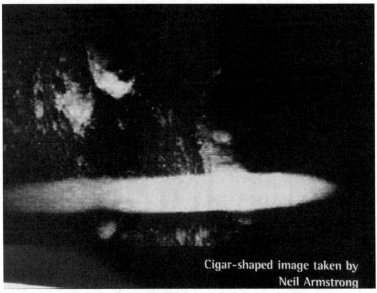

An image of a cigar-shaped object taken by Neil Armstrong during the Apollo 11 mission. (NASA)

Spacecraft on the Surface, Fleets of UFOs and Fast walkers

There have been several reports of people on Earth witnessing what they describe as cylinder-shaped spacecraft in the skies. Interestingly, cylinder-shaped craft have been photographed on and around the Moon. A NASA photo from *Apollo 16* (NASA No. 16-19238) shows a cylinder-shaped craft thought to possibly be a mothership, and estimated to be approximately two miles long. On the photo it is clearly seen hovering above the lunar surface. Cylinder-shaped ships are a theme that has been seen time and again when it comes to the Moon. George Adamski described several of the craft he witnessed as being cylinder-shaped (also sometimes referenced as cigar-shaped). Could these craft be the objects spotted by lunar watchers, lighted and sitting on the lunar surface?

Spacecraft sitting on the lunar surface has become a common theme in these times. Several lunar researchers are posting online videos showing what they believe are lighted ships sitting on the edge of craters, and speculate that some craters are being used as repair areas and hangeas for ships. There are also older NASA images showing what appear to be spacecraft parked on the

Moon. There is a legendary story of the *Apollo 11* astronauts having encountered ships sitting on the rim of a crater (covered in Chapter Nine). In one case, there is what appears to be an ancient spacecraft sitting on the Moon that some believe crashed. It can be seen in an image from NASA's *Apollo 15* mission (NASA photo No. AS20-1020). In yet another picture, there appears to be a large cigar-shaped spaceship, approximately nine miles long sitting on the Manilius impact crater. The image was taken by the Lunar Orbiter.

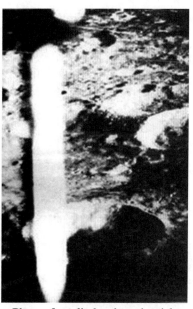

Photo of a cylinder-shaped aerial object taken during the *Apollo 16* mission. (NASA)

It could possibly be one of the craft of the same type that the *Apollo 11* crew are said to have witnessed sitting on the rim of a crater when they landed on the Moon. One wonders if these very same craft are the ones that so many have encountered here on Earth. Might any of them be the ones that followed the Apollo astronauts? There are fleets of spacecraft on the Moon. Several have been captured on videos and photographs flying in formation. There are a few Moon watchers who have observed fleets of UFOs leaving the Moon. One photographer was privy to seeing what appeared to be a fleet of UFOs exiting the Moon, presumably heading to Earth for more observation.

These fleets have appeared moving outwards and swiftly from the Moon. These could be the "Fastwalkers," that have been observed in the skies around Earth. Fastwalkers is the name used for what appear to be spacecraft that advance from deep space into Earth's vicinity. They are described as maneuvering and flying about in ways that human craft are not technologically able to do. The word "fastwalker" was purportedly created by the North American Air Defense Command (NORAD). These UFOs have been seen traveling at great speeds across the Moon as well. Several have been caught on video. NORAD is said to observe these objects closely from an underground facility located in Colorado. Whether or not these spacecraft approaching from deep space are connected to both the Earth and Moon is a mystery. Could they be coming from a mothership and here to study the Moon and Earth? Do they have operatives here? What would be their purpose?

Objects Crossing the Moon

Large, fast-moving objects traversing the lunar surface have been seen by Moon observers as far back as the 1800s. Some are estimated to have been traveling as fast as 6,000 miles per hour. There is no logical explanation as to what these objects are. In 1820, during a lunar eclipse, French astronomers witnessed unknown objects crossing the Moon in straight, even rows, and moving away from the lunar floor. Likewise, in 1869, three moon observers from the United States witnessed several objects traveling across the Moon in straight, parallel lines in what appeared to be an intelligent and methodical formation. In 1874, an astronomer from France stated that he had seen several dark objects crossing the lunar floor. Additionally, an astronomer from Czechoslovakia in that same year spotted a brightly lit object slowly moving across the Moon. In that case, the object eventually flew off into space. Could what these astronomers witnessed in these cases have been fleets of ships preparing for takeoff? Or, could they have been a type of lunar transportation by and for beings on the Moon, that has the capability to visit Earth and other areas of the galaxy?

In 1892, a Dutch astronomer noticed a dark spherical object moving horizontally as it travelled across the Moon. In 1896, William R. Brooks, an

Foto NASA # AS - 12 - 50 - 7346

An image of a UFO taken during the *Apollo 12* mission. (NASA)

astronomer at the Smith Observatory in America witnessed a black orbicular object also crossing the Moon. He stated that the object was one-thirtieth of the Moon's diameter. He described the object as moving swiftly, crossing the Moon in roughly three or four seconds! In 1899, two astronomers in Arizona reported seeing a brightly lit object traveling close to the surface. In 1912, an

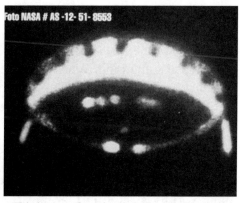

Foto NASA # AS -12- 51- 8553

This image of a spherical-shaped object was taken while *Apollo 11* astronauts were travelling to the lunar surface. It is estimated to be several hundred feet in diameter.

English astronomer by the name of F.B. Harris reported seeing an enormous dark object measured at 250 miles long and 50 miles wide on the Moon. Harris later stated that what he had seen could have been a shadow of something huge crossing above the Moon! Could it have been a mothership hovering above, like the one reported by Sergeant Willard Wannail? Of this account, Harold T. Wilkins wrote in his book *Flying Saucers on the Attack* (page 219), "It was like a crow poised, and I think a very interesting and curious phenomenon occurred on that night!"

William Pickering (1858–1938) was an American astronomer who made several noteworthy Moon observations. He was also well known for his creating several observatories including the prominent Percival Lowell's Flagstaff Observatory in Flagstaff Arizona. It was Pickering that discovered Phoebe, Saturn's ninth moon on March 18, 1899. One especially amazing observation that he reported was what he described as "traveling dark objects" that were travelling across the surface of the Moon. He wrote of this odd occurrence stating, "In trying to find conclusive arguments for or against the existence of animal life upon the Moon, I have necessarily studied not only the routes along which it appears to travel, but also the reasons for which it might be expected to travel."

Pickering determined that what he had seen was a form of

lunar insect! He reported that they covered 20 miles in 12 days. In his book *Our Mysterious Spaceship Moon,* Ufologist, Don Wilson writes, "No one questions Pickering's integrity and competence. He did see something. What it was remains a mystery." What did Pickering observe, and where is the phenomenon now? *Is this activity still occurring on the Moon today?*

The Kareeta

If we look at the tales of America's Old West, there are stories of extraterrestrial contact. Out of Texas comes a tale about an otherworldly spaceship that crashed there in 1897. Purportedly, the being piloting it died and was buried in a cemetery there. Could that being have come from the Moon? Even more weird is the story out of Tombstone, Arizona, where it is said that during a gunfight at the OK Corral, the men allegedly fired on an enormous metallic bird. Although that metallic bird could have been an alien craft resembling a bird or even an airplane, it brings to mind the story of the Kareeta, an extraterrestrial spacecraft seen one year in San Diego, California. As the story goes, an object referred to as a "mechanical bird" was seen flying across the Moon on October 9, 1946 by residents of San Diego during a meteor shower. Purportedly, the name "Kareeta" was telepathically given to Mark Probert (a psychic) from benevolent beings aboard the craft. Probert along with eleven other people witnessed the strange object.

An article was written about the Kareeta in the *Round Robin* journal in 1946. Whether or not people believe in psychics is

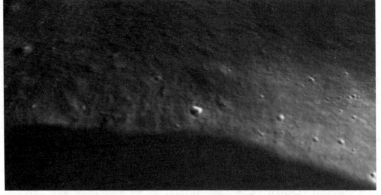

This triangular anomaly with rows of seven light-like dots along its edge has been tracked down on the Google Moon viewer.

irrelevant in this case, because the article in the *Round Robin* cites several other witnesses who testified to seeing this "mechanical bird." Could it have been the same type of alien aircraft as the one in Tombstone? It seems like an odd coincidence. Spacecraft shaped as birds are also mentioned elsewhere. During the *Apollo 7* mission, the astronauts photographed a UFO that looked similar to the shape of a bird. Also, there are bird-like flying craft mentioned in ancient Sanskrit writings. However, this is just one of many designs of spacecraft seen flying near the Moon and Earth.

The Triangular Anomaly

The "triangular anomaly" is an object located on the lunar floor that has been recently scrutinized. Images of it have made their way around the internet. It is a "gigantic" construction in the shape of a triangle and is believed to be either an enormous spacecraft or an extraterrestrial moon base. It is so large that one source even suggested that it may be a spaceport used for stowing and dispatching alien spacecraft. Images of the triangular anomaly show seven straight rows of what appear to be lights running alongside of the edges of the construction. In an article about the construction, the Altered Dimensions website had this to say: "It's too symmetrical to be a random topographical structure. It's not faked, it's really there, and looks nothing like the craters that surround it." The Tech and Gadget News website also had comments about the object stating, "The gigantic shape actually looks like the leading edge on an immense, triangular spaceship, similar to, so far, super-secret stealth aircraft technology, but is much larger than any airplane ever built on Earth." What could

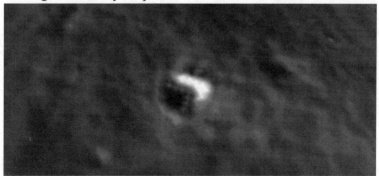

Close-up of the triangular anomaly with rows of seven light-like dots along its edge as tracked down on the Google Moon viewer.

this be? For certain it is there, it is real, and it should be monitored. The coordinates on Google Moon for the triangular anomaly are 22042'38.46N and 142034'44.52E.

More Lunar Anomalies

Mysterious lights are not the only strange phenomena happening with the Moon. All of the activity, of course, may not be extraterrestrial related. However, there are many unexplainable events that happen. Much of the time, these phenomena go unnoticed, and people have no idea that they even occur.

The Black Line Phenomenon

One of the most perplexing phenomena seen in recent years is the "Black Line Phenomenon." It involves a mysterious black line (or lines, there are sometimes two) moving vertically across the Moon. People have observed this phenomenon for over 30 years, in various locations around the world. It is seen most often during a full or super moon. One witness reported seeing the line during a super moon as he was driving along a highway in the northern UK. He reported that it resembled a line drawn with a black marker. The line has been variously described as sharp, vivid, and remaining in a consistent position as it passes across the Moon. There have been several speculations as to what it could be. Theories include it being an atmospheric phenomenon, an optical illusion, and the work of extraterrestrials. Some have even suggested that this may be the same phenomenon that is found in the Koran, when the Prophet Mohammed performs the miracle of the "splitting of the moon."

The most prevalent theory is that the line is from contrails. Some observers, however, disagree with the contrail hypothesis. They maintain that contrails have jagged edges, are extremely volatile, break apart, dissipate and at times change position. This odd black line has none of these characteristics, as it remains uniform and stays within the Moon's edges. Could this weird phenomenon be a sign from extraterrestrials that they are on the Moon? Are they attempting to signal us via an obvious black line?

If it is a sign, then it is one that has caught people's attention. It certainly has us taking notice and asking, "Who or what is causing that effect?" Similarly, there have been odd lines seen in some of the Moon's craters. Although there is probably no relation to the one crossing the Moon, it is certainly warranted to mention them. On the night of July 6, 1954, astronomer Frank Halstead observed an even, straight black line in the Piccolomini crater. Piccolomini had been monitored in the past, yet nothing like what Halstead reported had ever been seen. Several other people witnessed the unusual line as well. Eventually, the line vanished just as mysteriously as it had appeared. Another puzzling account comes from the Eudoxus crater. In the late 1800s, several moon watchers witnessed what looked to be a glowing cable line slowly stretching across Eudoxus from west to east, until it covered the entire crater. They watched for nearly an hour before it flickered out. What could have been the meaning behind the glowing cable line? Is it connected to the straight line crossing the Moon? Is it a signal to Earth? Perhaps it was simply lunar inhabitants going about their business.

Who's Messing with the Moon?

The Moon appears to be changing. That is according to Moon observers past and present who assert that there have been peculiar, unexplained changes that have taken place on the Moon for decades. Astronomers in the past chronicled changes on the lunar surface. Today, there are those who claim that the face of the Moon is changing, as if there is someone on the Moon making modifications that alter its appearance. Some claim that in the past few years the color of the Moon has subtly changed. Some have observed that in some parts of the world the Moon is appearing to be more of an oval

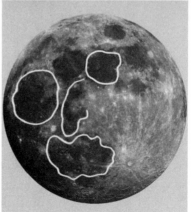

Is the Man on the Moon being destroyed before our eyes?

shape, and it seems larger.

People have even alleged that there is no longer a "man in the Moon" during the crescent Moon. It is claimed that the nose and the mouth have disappeared. Is this correct? Are some people seeing the Moon differently than others? Is the Moon really changing? If so, then who or what is causing it and for what purpose? George Leonard too mentioned this in his popular book *Somebody Else is On the Moon* (page 23) in which he asserts that the Moon is occupied, stating, "They are changing its [Moon] face."

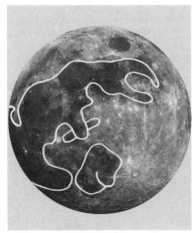

The perceived Rabbit in the Moon mainly seen in China.

I recently came across a rather unusual theory. A lunar researcher is convinced that the Moon is getting a "facelift." This researcher states that the Moon is so old that the Moon's creators are upgrading it. She researcher asserts that what appears to be a change in color, the changing face of the Moon, and other indications of the Moon being altered are caused by extraterrestrials that oversee the Moon. The theory is that they are renewing it. Could she be correct? Could our Moon be maintained by extraterrestrial overseers that are charged with the upkeep, repair and care of it? Would this be due to it possibly being an artificial satellite? The idea of the Moon having maintenance may seem farfetched to some, but who is to say that there are not caretakers in charge of the Moon? There could be those that are tasked with ensuring that the Moon does not "malfunction." That would be a disaster for life on Earth.

There is the theory that the extraterrestrial group known as the Anunnaki brought the Moon here and own the Moon. Some believe that it is Anunnaki ships that some astronomers see entering and exiting the Moon. Could the Anunnaki believe that it is time to renovate the Moon in some way? If so, then how does that affect the Earth? If the Moon is artificial, then I would wager that any type of technology would need an overhaul. If there is a lunar overhaul in the process, we should ask if this has happened before.

Is there a maintenance schedule in place that occurs at specific periods of time?

Perhaps we did not notice it before because a prior upgrade may have occurred before the age of the telescope, and we just did not know about it or understand what we were looking at. Remember, there was a time when humans thought the Moon was a god. There was a time when a lunar eclipse was thought to be a sign of evil and a curse. Any changes to the Moon before our current time may have gone unnoticed. An upgrade to an artificial entity is not out of the realm of possibility.

A Holographic Moon

Perhaps one of the strangest topics surrounding the Moon that has come about in the past few years is that there is a holographic image covering the face of the Moon to conceal what is on the lunar surface. Many believe that there are Moon cities being covered up via this hologram. A strange event with the Moon happened when a lunar observer was looking through his telescope at the Endymion crater. There the observer claimed to have seen a Moon city. This story set the stage for a lot of conversation and theories as to what could be happening with the Moon. This day, as the observer was studying the Moon, he stated that for a moment, he saw a city. He insinuated that for an instant he saw the city, and suddenly it was gone. He believed that somehow, there was some sort of glitch in a holographic system that caused the projected image of the Moon to come down, revealing what was really behind the proverbial, "veil." It has been speculated that what was witnessed that day was a hologram collapsing just long enough for Moon observers to see the truth.

Contactees and the Moon

Earlier, I discussed George Adamski, the man who claimed to have been approached by a Nordic extraterrestrial being. There are several other people that claim to have had contact with beings from other worlds. They are referred to as "contactees." What is most interesting about so many of these people is that their claims line up with information often discovered in the future. Therefore, even though many people do not believe in such accounts and are even baffled at the mere idea that there are people in contact with extraterrestrials, we cannot discount what they are telling us. Who is to say that contact with extraterrestrials is an impossibility?

Of course, as in all things, caution should be used. However, many people have laid their reputations on the line to report what they have seen and witnessed by way of experiences with otherworldly people. It is time that we sit up and listen, because there just may come a time when they are bringing life changing information sent from extraterrestrials trying to assist. When it comes to contactees, there are those who allegedly have met space people, interacted with them, and have even been taken aboard spaceships and traveled into outer space. Some have stated that they visited the Moon. What they claim to have seen on the Moon is astonishing.

In the *Encyclopedia of Moon Mysteries, Secrets, Conspiracy Theories, Anomalies, Extraterrestrials and More,* contactees are defined as:

> Someone that is in communication with extraterrestrials. Such a person is generally in contact with the extraterrestrials through telepathy, channeling and in some cases, direct visitation. The main reasons extraterrestrials give for approaching a contactee, is to assist mankind and to prevent humans from using nuclear warheads. The extraterrestrials have stated to some that they are here to monitor what is being done in regard to nuclear war, to prevent mankind from damaging the Earth and also the Moon. Some contactees have reported boarding a spaceship and traveling with extraterrestrials to a remote place or another world and in some cases the Moon.

An extraordinary tale comes from ufologist and Author George Washington Van Tassel, (1910–1978). His books include *I Rode a Flying Saucer, Into This World and Out Again, The Council of Seven Lights, Religion and Science Merged* and *When Stars Look Down.* Van Tassel claimed to have been first visited by a Nordic extraterrestrial when he was in Palm Springs, California in 1952. This contact occurred at 2:00 in the morning. He recalled this experience in his book *Into This World and Out Again.* He was visited a second time on August 2, 1953. Van Tassel stated that the extraterrestrials travelled on a 36-foot-long bell-shaped antigravity scout spacecraft. Van Tassel said that this craft was one of the smaller ones. Scout ships were sent out from a larger vessel. Interestingly, this larger carrier would visit the Moon for provisions at a lunar station that had been used, he said, for thousands of years.

New Zealander Alex Collier says he is a contactee for the Andromedans. They are a race of humanoid beings from the Andromeda galaxy. Collier's main contacts included two Andromedans by the names of Morenae and Vissaeus. The extraterrestrials purportedly taught Collier a great deal about galactic history, including that of the Moon. He was given a considerable amount of knowledge about the cosmos that he has shared through his writings and lectures. He covers such topics as extraterrestrials (including existing races on Earth, advanced beings, and the Reptilians), the dark forces, Mars, the new world order and more. Collier, who is still active today, has been informing the public about extraterrestrials for years and has shared quite a bit of information received from them regarding the Moon.

In a lecture given in 1996 titled, "Moon and Mars," Collier stated that on the far side of the Moon, there is an atmosphere. During an interview in 1994, Collier discussed several Moon topics. For one, Collier stated, that extraterrestrials brought the Moon into Earth's orbit and that the entities that performed this task exist among us. Collier suggests that there are areas of the Moon that thrive with vegetation. He also believes that the Moon has water. Collier once stated, "Our Moon has an atmosphere that is in many respects similar to that of the Earth. In many large craters on the visible and the invisible hand, the atmosphere is denser than sea level on Earth, it is claimed." Collier also claims to have received information from the Andromedans about how the Moon came to be in Earth's orbit. He explained that the Moon had been placed in the tail of a comet and dragged into our Solar System.

This information is quite like that of the Zulu tribe that tells of the Moon being sent across the universe to Earth. It is also similar to the theory of the two Soviet scientists that wrote that the Moon is an artificial satellite, created elsewhere and then brought here. In his Moon and Mars lecture in 1996, Collier made the following comments. "I'll tell you what the Andromedans have said about the moon. They have said that our moon is an artificial satellite, in fact it is a spacecraft. Much of the debris on the surface was put there and was built purposely to make it look like what it isn't. OK, it is hollow, it is metal underneath it, and it has the ability to leave our orbit under its own power. They say it came from Ursa Minor." He continues with, "It was built (this sounds just like Star Wars) around that 17th planet. It was then put into the tail of a comet and then dragged here."

Lunacy

One of the strangest claims connected to the Moon is that it can cause "lunacy" in some people. Being a lunatic is a term used for a psychologically disturbed individual. The word "lunatic" is derived from the Latin word "Luna," which means moon. People once believed that a person engaging in rash or psychopathic behavior had been affected by either a full Moon or the phases of the Moon. Some people still belief this. However, there are scientists that argue that this area falls into the realm of myth.

The word "moonstruck" has a similar meaning. When a person is said to be "moonstruck" it implies that the person is exhibiting bizarre, crazy, or unpredictable behavior (lunacy). Also, the term "Loony bin" (an old-fashioned term for an institution for the mentally ill) came from the Latin word Luna. In the book titled *Nothing In This Book is True, But It's Exactly How Things Are* (page 183) by Bob Frissell, it states, "The day before, the day of, and the day after the full moon there is usually an increase in murder, rape, and other crimes because the moon causes a bubble in the magnetic field. This minute bubble is enough to push already emotionally disturbed people over the edge."

Could this potentially be one of the reasons the world is out of sorts at times? Not withstanding all the benefits of our life sustaining moon, could it be on some level a part of the problems with the craziness seen on Earth over the past centuries? Perhaps this was an oversight on behalf of the creators.

Death and Rebirth

In ancient Hindu belief, it was believed that the souls of those who died departed Earth and travelled to the Moon to await rebirth. There were also some religious groups of ancient Greece that believed the Moon to be the abode of those that died. The Tatar people from Central Asia referred to the Moon as the "Queen of Life and Death." Today, there is a strange story that has circulated for years that states that the Moon is a "reincarnation machine." The implication is that it is where the souls of the dead go. The light of the Moon is said to attract them. There they are recycled and sent back into a different incarnation on Earth.

There is even a story of a remote viewer who followed the soul of a person that had just died. He tracked the soul to the Moon. As they approached a structure on the Moon, the viewer noticed that there were warnings not to come near a particular object that was square shaped. This object sat on top of a tower. The square-shaped object is believed to have been a machine that caught souls. There the souls are said to be recycled in this "prison" tower.

(Author's note: This is a very disturbing story. As a writer on topics involving life after death, I would like to say that this is not the place where souls end up. Of course, there are many paths in the universe, and if there is any truth to this very strange tale, then the trapping of a soul would be by way of either a fluke or trickery. It is not our destiny when we die. After death, people are advancing to other dimensions, worlds, and returning to their original home after death. A soul machine is not where everyone goes… if there is any truth at all to this tale at all.)

A Ranger 7 photograph of some curious craters with dome-like structures in them.

Chapter Six

A Moon Metropolis!

Many phenomena observed on the lunar surface appear to have been devised by intelligent beings.
—Ivan T. Sanderson

Was there once an ancient city on the Moon? Is there a city on the Moon today? Some lunar researchers believe the answer to both of those questions is a resounding yes! If someone is residing on the Moon then, what is the evidence for this claim? If there are lunar inhabitants, they seem to have been on the Moon from the beginning. No matter if they were in a "spaceship Moon," or if the Moon were seeded by extraterrestrials that located the Moon and started life there, it appears they have always existed. Whoever it was lived there a long time ago and may still be there now. There is a theory that there could have been a time when the Moon had an atmosphere and water. No matter how implausible that may seem, we have no idea how an extraterrestrial species would have been able to exist. However, something or someone did, and whoever they were may have had an important Earthly connection.

In her book *ET's are on the Moon & Mars*, C.L. Turnage explains what the astronauts witnessed while on the Moon, stating, "In the winter of the 1974 issue of *UFO Report*, an analysis was made of astronaut voice tapes. These tapes revealed the startling observations of the perplexed spacemen upon seeing what appeared to them to be unusual, and unnatural formations, resembling the ruins of a dead civilization." In the years following, researchers have spent a lot of time examining images taken during the Apollo missions and those from the *Clementine* mission. Numerous objects were discovered in the photographs that they alleged were ruins and artifacts (some partial and some complete). Ruins purportedly

121

found on the Moon include monuments, partially destroyed walls, complex geometric-shaped constructions, structures, pyramids, obelisks, monoliths, terraces, tunnels, and other objects that do not appear to be there by natural geological means.

Some of the structures appear to have suffered some sort of destruction. In his book *Extraterrestrial Archaeology* (page 60), researcher and author David Hatcher Childress features an illustration showing a "layered outcrop of rock" that is approximately "8 meters tall" located close to where *Apollo 15* sat down. Childress writes, "It's similar to the polygonal building style used at the massive wall of Sacsayhuaman in the Andes of Peru. Walls like this on Earth can be artificial." He asks the thought-provoking question, "Can they on the Moon as well?" What does this imply for our history and our life today? Does it affect us? It may! Because there is a chance that the ruins on the Moon may be directly connected to Earth.

We may be galactically and historically tied to the Moon (beyond what the Moon does for Earth in nature) without even realizing it. Additional objects discovered in the images included roads, pipelines, domed craters, spacecraft, rows of buildings, a plane-shaped UFO, a cylindrical-shaped UFO, a landing pad, satellite dish, and perhaps others that we do not even recognize as alien structures! Former NASA expert geologist Farouk El Baz is quoted in a 1974 article *in SAGA Magazine* (Volume 47, Number 6, March 1974) stating about objects on the Moon, "We may be looking at artifacts from extraterrestrial visitors without recognizing them." Certainly, there may be artifacts placed there by lunar inhabitants that we do not reconize.

Many of the objects located in the photographs appeared to be structures from an ancient city. The scientists, researchers and others were shocked to learn that we had found evidence of intelligent, extraterrestrial life! It was startling! And it was classified. In fact, it is interesting to note that we did not get further than our closest neighbor in the solar system before we encountered evidence that we are not alone in the universe. If we have found evidence that there are other intelligent beings out there besides ourselves, and on the Moon, then we must ask ourselves, "Is this an indication that the entire galaxy is made up of life?" Is it like the old adage,

"Where there is one, there are many?" Is what we discovered on the Moon just a drop in the bucket as to what or who is out there in the universe? It appears that our trips to the Moon took us down the proverbial "rabbit hole." We have walked through a door and found ourselves in another place, and it has changed our reality. I might add that we do not need several structures to show evidence of other life out there. We only need *one* to tell us that someone else was there, and we are not alone. So then, we must consider if there is even more going on out there besides what was found on the Moon. How many other extraterrestrial groups might there be? *What else don't we know?*

This is truly a case where we cannot unsee what we have seen. *Is ignorance bliss?* For some perhaps. Most interesting is the prediction by an Austrian space engineer that worked with NASA from 1959-1974 by the name of Josef Franz Blumrich (1913–2002). The author of several books, Blumrich is best known for his iconic work *The Spaceships of Ezekiel* (1973). After investigating the subject of extraterrestrials for over a year and a half, Blumrich concluded that they had consistently visited both the Moon and the Earth over time. In 1974 Blumrich contributed a synopsis of his ideas to the *UNESCO Science Report* (a global monitoring report published by the United Nations Educational, Scientific and Cultural Organization) titled, "Impact of Science on Society." He was certain that in the near future, we would discover artifacts left on the Moon by extraterrestrials. The question is, "Are these beings still around, living in or near this ancient city?" If not, then where did the inhabitants go? Are they still there in a more modern version of their city? Or did they go underground due to the harsh elements on the lunar surface?

It appears that this very important, mind-altering discovery of an ancient civilization on the Moon threw everything off when it came to the Apollo missions. It changed people. It caused cover-ups. The knowledge of it altered the perception of those who knew. For others, the secret was too much to bear. As a result, there are tales of emotional breakdowns. Others were outraged that the information was not covered in the public forums and moved forward with breaking disclosure agreements and sharing with the world what they knew. Missions to the Moon ceased, with the official story being that the cost of Moon missions were exceedingly high, and that the public had lost interest in the lunar missions entirely.

It is the theory among some that the real reason for the missions to the Moon, and for the space race in general, was to look for remnants of this ancient civilization. It appears that there was some knowledge of anomalous structures there from the beginning. There certainly was knowledge of the anomalous lights. If that is not the case, then we can only imagine the surprise of the astronauts who traveled to the Moon to locate sites for a human colony and ended up discovering the ruins of a past civilization instead. Is this the reason that none of the world governments have placed humans on the Moon? It does appear that once mankind finally advanced enough to travel into space and headed towards our closest cosmic neighbor, we found something unexpected out there and it was right next door. There could be a worry as to what else we might discover. No matter how much people say they want to explore the universe and look for other life out there, it appears that for many this is not true. I don't know that people really want to know the truth as to what is out there. It is possible that the dream of seeking out new worlds and new civilizations ended with the possibility of actually achieving that goal. Was it too soon?

In the 1960s, when we achieved space travel, the public was excited about landing on the Moon. Finding beings in outer space was not their focus, and the masses may not have been prepared for it at that time. During that period most people thought if there were intelligent life out there, it would be discovered years into the future. Others believed that if life existed, it would be light years away, and not so close to us. This even though the evidence shows that Earth has been visited since the beginning. There were officials who knew there was a possibility of encountering life, as NASA commissioned a study known as the Brookings Report on the subject to help them to understand how to proceed with the public if that happened. The conclusion was that the world was not ready. It was concealed, as best it could be. As a result, many of the stories of the Apollo missions have been relegated to the area of tales and fables. However, the activity on and around the Moon is still happening. And Earth's skies are filled with anomalous sights that officials are finding increasingly difficult to explain away as they just may be coming from our Moon (and elsewhere, but that is another book). *Are people ready to know now?*

When the Apollo astronauts returned from the Moon in the 1960 and 1970s, there was no shortage of speculation about the photographs they took during their missions. Numerous books have been written about what they saw and experienced there. Lunar researchers studying photographs of the Moon stated that some of the structures appear to have been abandoned, while others appear to be in partial ruins. This has led them to believe that a catastrophic or tragic event may have taken place on the Moon eons ago that either annihilated the lunar inhabitants, or sent them fleeing off the Moon in ships, or sent them underground. There are what appear to be vitrified and melted ruins which tells us that quite possibly a war took place there. Most are partial structures, and there is debris which appears to have been caused by explosions. According to researchers, these ruins appear to have once been prominent structures. The remnants of this city could be a billion years old. Since the Moon does not have an atmosphere, the ruins remain there, as there is no rotting, decaying or corrosion. Some researchers speculate that these beings may still occupy the Moon today. Perhaps this is why the Moon seems to have both ancient ruins and modern buildings. If there were a war on the Moon that obliterated a civilization, then it could be that a new race of beings now exists there. *If so, are they friendly?*

The discovery of structures on the Moon was eventually presented in a public forum on March 21, 1996, at the Washington National Press Club. An excerpt from the official press release states:

> NASA scientists and engineers participating in exploration of Mars and Moon reported results of their discoveries at a briefing at the Washington National Press Club on March 21, 1996. It was announced for the first time that man-caused structures and objects had been discovered on the Moon." Scientists at the function maintained that the constructions were being investigated, and results would be released at a later date.

Some highlights of the meeting included:

*The Soviet Union having photographs that could prove the presence of extraterrestrial occupation on the Moon.

*Numerous photographs revealing several areas of the lunar surface where evidence of extraterrestrial activity was apparent.

*Photographs and film footage taken by NASA astronauts.

During the meeting, consultants answered questions as to why the public had not been told of the findings. They were quoted as saying, "It was difficult to foresee reactions of the people hearing the news that extraterrestrial creatures were or are still on the Moon." Since that time there has been a great deal of study and speculation when it comes to these findings. According to sources, the ruins extend across the lunar surface for miles. The structures appear to have been crafted by a highly developed and intelligent race of beings. Additionally, if there is a current civilization on the Moon, there may well be both ruins and modern buildings and bases created by extraterrestrials. We should remember too that the Apollo astronauts have explored less than one percent of the lunar surface. There are still plenty of areas on the Moon to be investigated. Once we return to the Moon and are able to cover more of the surface, we just may find what's really going on there.

Cities on the Moon

Astronomer Franz von Paula Gruithuisen (1774–1852) was best known for his notable investigations and analyses of the Moon. He was certain that he had detected no less than an entire metropolis on the lunar surface. He called it *"Wallwerk."* The region in which Wallwerk was located was in a rocky, uneven area of the Moon north of the Schroeter crater. This area was fraught

Drawing of a Moon city by the Astronomer Gruithuesen.

with rectilinear ridges that resembled a fishbone pattern and geometric shapes.

In 1824, Gruithuisen publicly announced his discoveries in an article he wrote, titled "Discovery of Many Distinct Traces of Lunar Inhabitants, Especially of One of Their Colossal Buildings." His findings were later negated. It was thought that Gruithuisen had misidentified the lines he was seeing for streets, buildings, and canals. Today, we seem to have come full circle since the days of Gruithuisen's research, because for years afterward, the idea of there being a city on the Moon was considered ludicrous.

Then, NASA was created. Space probes and men were sent into space and to the

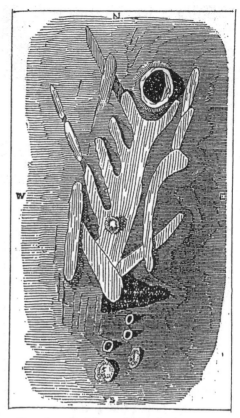

Drawing of a Moon city by the Astronomer Gaudibert.

Moon, and what we thought we knew changed in an instant. When we entered space, there was not only a dark, vast emptiness. There were UFOs and strange lights. There were people coming forward with tales of the astronauts seeing extraterrestrial ships. Earth skies were filled with strange objects eventually dubbed UFOs and now UAPs. Today, we have people from all over combing through pictures, transcripts, videos and more in an effort to understand what is happening on the Moon, and if there is any connection between there and Earth. It seems that Gruithuisen was not so far off in his assessment of the Moon after all.

We had been taught that due to the lack of atmosphere and the extreme temperatures on the Moon that nothing can survive there. Yet, there appears to be more evidence of a civilization once existing on the Moon than not. Given what has been discussed in the previous chapters (i.e., strange happenings, structures on the surface, lights appearing and disappearing, objects crossing the Moon and more), it certainly appears that it is quite busy there. If there is no life on the Moon, then where is all of this coming from? If there is evidence that there was or still is a civilization on the Moon, how are they living there in those dire, extreme conditions? There are no solid answers, only contradictions. It appears that the Moon even has the scientists stumped!

There are two hypothetical scenarios when it comes to there being a city on the Moon. There is a theory that there is a civilization with a city on the surface of the Moon. It is also thought that several craters may have habitats located inside them. There are also reports of a possible underground city. Which is correct? Evidence shows that there may be two different kinds of artificial environments happening there. Some may be located on the surface of the Moon, others within the interior.

The Lunar Surface

When the telescope was first invented, there were astronomers who supported the idea that the Moon is inhabited. There were those such as Johannes Hevelius (1611–1687), a renowned Polish intellectual and astronomer who in 1647 published an atlas of the Moon titled *Selenographia*. It included a comprehensive map of the lunar surface outlining the lunar topography and features. Hevelius argued that the Moon was inhabited by Selenites (the name for beings on the Moon). He even made it into NASA's "Chronological Catalog of Reported Lunar Events," where he is listed as observing something strange on the Moon in 1650. He described it as a "red hill" in an area around the Aristarchus crater. These types of sightings convinced him that there was a race of people residing on the Moon. But that was then. Where are we today on the subject?

We have people such as George Leonard (1921–) whose book *Somebody Else is on the Moon* (1977), has become a classic and is still being talked about today. George Leonard is a former NASA scientist and photo analyst that once worked in the photographic intelligence division of NASA. While working at NASA, Leonard was assigned the job of analyzing images taken on the Moon by space probes sent up in the sixties. During that period, NASA had been exploring ideas for safe landing sites for the Apollo missions, as well as charting the Moon. While examining the photographs, Leonard came across numerous images that contained anomalies on the lunar surface. In his book *Somebody Else is on the Moon,* Leonard published the results of his in-depth, painstaking examination of thousands of images of the Moon, where he located mysterious objects. Leonard claimed to have located large machinery, constructions in odd geometric shapes, towers, track marks, huge domes and more. It was his belief that these images show proof of extraterrestrials living on the Moon.

Johannes Hevelius and second wife Elisabeth observing the sky with a brass sextant (1673).

About his discoveries, in *Somebody Else is on the Moon* (page 23) Leonard writes, "The photograph, with others in my collection, fairly screamed out the evidence that the Moon has life on it. There was no denying the truth was shone through: the Moon is occupied by an intelligent race or races which probably moved in from outside the solar system. The Moon is firmly in the possession of these occupants. Evidence of Their presence is everywhere: on the surface, on the near side and hidden side, in the craters, on the Maria, and in the highlands."

Given what we have learned about all the anomalies on the Moon, we can at least speculate about the Moon being inhabited and even having a city of some type on the surface (even though to most this is a farfetched idea). In fact, if we listen to the research of some recent lunar researchers, there appears to be what may be new structures on the surface. No matter how we slice this, it comes up with a type of extraterrestrial metropolis on the Moon. Or at least whatever an extraterrestrial city would entail. Why haven't we discovered them outright? Why can't we see them? I will tell you what one radio host said to me when I was being interviewed on the subject. He stated, "They [the extraterrestrials] could simply turn the lights off."

That may be exactly what is happening. They may be hiding, not just from us, but other extraterrestrials as well (speculation here). There may be reasons for them to do so. However, every now and then, a little evidence comes in and we get a peek inside their world. As we have established in Chapter Five, Strange Happenings, there are lunar researchers that have seen lights inside of craters. We can only imagine where the lights are coming from, and what they mean. Could they be signs of a people living on the Moon simply going about their lives in a city or town that is located inside of those brilliantly lit craters? Some Moon watchers have claimed to see what looks like intelligently shaped objects hidden inside some of these lighted craters. Even more interestingly, some craters appear to have domes (some are said to be glass) covering them, while others seem to have a type of strange covering to hide what some believe may be structures underneath. There are even some that have witnessed UFOs over the craters. These are thought to be spaceships, or some sort of biological entities. If there really is a city, then it appears that this is a very high-tech city.

The Interior

In their book *Extraterrestrial Civilizations on Earth* (page 78), co-authors Cecelia Frances Page and Steve Omar write, "If you lived on the Moon, Phobos or Mercury, you would need an artificial, underground air-conditioned domed city totally protected from outside atmosphere, or a body human in shape yet made out of pure energy." Could this be what we are looking at on the Moon? Especially with the reports of lighted domes covering craters. Is an underground settlement not something that we would consider planning ourselves? The domes seen over craters are speculated to be a part of a large underground system. These domes appear to be lighted at times. Is there an artificial environment beneath these domes? We have already looked at the hollow moon spaceship theory. However, to take it one step further, we could consider how beings may be living within the Moon's interior.

One of the first thoughts that comes to mind is that the Moon is a megastructure. *The American Heritage Dictionary*, defines a megastructure as, "a very large building, often containing modular units, designed to allow a community to be self-enclosed or self-sustaining." Another theory is that caverns inside the Moon were converted into living areas. In Jim Marrs' book, *Above Top Secret* (pg. 228), prominent geologist Farouk El Baz a former secretary of the landing site selection committee for NASA's Apollo Program, is quoted as saying, "There are many undiscovered caverns suspected to exist beneath the surface of the Moon." In fact, what appears to be openings to several tunnels, have been spotted on the lunar surface. Could these be entryways to an artificial living environment below? Or an underground metropolis of sorts? Mysteriously, the *Apollo 11* astronauts alluded to beings living inside of the craters. The following is an excerpt from an *Apollo 11* conversation.

CMP: Boy, there must be nothing more desolate than to be inside some of these craters, these conical ones.

CDR: People that live in there probably never get out.

Contactee Alex Collier once spoke of what he learned from his extraterrestrial associates about the Moon. In the article "ET Contactee Says Moon Has Vegetation and an Atmosphere," Collier is quoted as saying that on the Moon there are "huge underground facilities [that] contain large lakes, plant life of Earth, food warehouses and hangars for spaceships with alien written texts on the walls in the portals." Paracelsus C is an impact crater located on the dark side of the Moon. In images taken by *Apollo 15* astronauts and the *Lunar Reconnaissance Orbiter*, there are two constructions in Paracelsus C that have been found by lunar researchers. They appear to be walls positioned on the crater

An image of the far side (dark side) of the Moon. Exterriastrial bases are believed to be located there.

floor. After careful study of the pictures, researchers maintain that they are walls that look as though they form a passageway. It is believed that this passageway leads beneath the Moon's surface.

Of course, our main ideas about extraterrestrials are based on what we see in science fiction books and movies, etc. When picturing a city on the Moon, we imagine humanoid inhabitants living there. It is fascinating to let the mind wander and imagine what it would be like for extraterrestrials living on the Moon. We have these images from NASA of the Moon that portrays it as white and grey, dry and even flat. Buzz Aldrin referred to it as having the appearance of "magnificent desolation." However, the astronauts visited very limited areas of the lunar surface. In reality, the Moon also has color, mountains, hills and a varied topography. Just as we gaze up at the Moon in awe and wonder, extraterrestrials too may look at the Earth and never imagine that it is a populated bustling world.

If there is a city on the Moon, then the Moon populace may be just as oblivious to us as most people on Earth are to the idea that there may be extraterrestrials and structures on the Moon. If there really are Moon inhabitants, they may not know that for decades now, people on Earth have been studying the Moon and trying to determine what is going on there. Then of course, they may be a well-versed populace and not as in the dark about the universe as we are. Some have speculated that if they are up there, they may know about us and want nothing to do with us. Or they may know nothing of us as their governments, just as ours, may be secretive as well. The general populace there may be in the dark. This is especially true if the city is beneath the Moon's surface. This is all speculation of course. But, how exciting! If we want answers, we need to look at the evidence, ask the questions and get a conversation started about what really is happening on the Moon.

Once we return to the lunar surface, we will see for ourselves if the Moon is inhabited, and if there is a city or some sort of habitat either on or beneath the surface. There is a possibility, too, that a city might be camouflaged to match the environment, to go undetected. A case in point would be the experience of the *Apollo 17* astronauts where they came across a structure on the Moon that looked as though it was made from the same materials as those of the lunar surface. The color was the same. (Details of this event can be found in Chapter Nine.) In other words, it seemed to be camouflaged to fit into the landscape. If this very interesting account is any indication of what to expect from a least one city located on the Moon, then we can assume that it or they are designed in a way to go virtually undetected. This of course may not be due to any purposeful deception on the extraterrestrials' part; it may just be an architectural style. However, we can only wonder if they are attempting to conceal themselves from the people of Earth, or other alien species in general, as this architectural style would be difficult to spot from a distance.

In his videos, Astronomer and Lunar Researcher Bruce Swartz of *Bruce Sees All* (YouTube), has gained a large following for his studies of the Moon. He has posted some very captivating videos on the subject of Moon cities. He has spent countless hours attempting to film and show the audience what he has discovered. According to Swartz, he has allegedly found several structures on the Moon, and he is not alone. With today's technology, there are others coming forward with reports of what appear to be modern artificial constructions on the lunar surface. These are outside of the original images of ruins on the Moon that appeared in Apollo photographs in the 1960s and 1970s. These are something *else*. Swartz claims to have located structures that are symmetrical, geometrical, and full-on buildings with even edges. He has described seeing towering structures, and a possible concealed city.

Researcher Scott Warring has done some amazing work in uncovering lunar anomalies. In the April 30, 2020 article on the website brobible titled, "UFO Expert Says He Has '100% Indisputable Proof' Alien Cities Exist On The Earth's Moon," by Douglas Charles, Warring speaks of a type of structure that he located on the lunar surface. Warring states, "These are flat black, meaning they have no reflective surfaces. They are massive, up to hundreds of kilometers long. They often have the small buildings near their entrances with as many (4-8 usually) like attachments going back and forth." If Swartz and Warring are correct in their assessments, we can quite rightly investigate the topic of a modern-day moon metropolis more confidently. There will certainly be more to discuss in the coming years.

A Domed City

Perhaps one of the most attention-grabbing discoveries on the Moon has been the domes. These enormous glass dome structures have been reported on by astronomers for centuries. German astronomer Johann Schroeter who observed domes on the Moon, believed they were created by lunar inhabitants. In the early 1900s, French astronomers went public with their findings of domes on the Moon. In the first modern-day discovery. a dome was located in images taken in the region of Sinus Medii by the Surveyor 6 lunar lander. One image shows a dome positioned above the rim of the Copernicus crater. It has been described as translucent and

An artist's concept of a domed city on the Moon.

radiating a bright bluish light that is emanating from the dome's interior. Lighted domes such as this have led some to surmise that they are the result of superior technology, with a great deal of power behind it. It is thought also that there is major activity going on beneath the domes. *Perhaps a city below?* At one point, a cluster of 20 to 30 domes was observed in Tycho crater. In 1966, NASA released 33 Moon dome images that were taken by the *Lunar Orbiter 2,* space probe. NASA's Ranger 2 probe (launched on November 18, 1961) purportedly took over 200 photographs of craters with domes located within the interiors. China's Chang'e-3 space probe photographed what is believed by some researchers to be ancient shattered domes. Some researchers and ufologists believe the domes were created by extraterrestrial colonists that created an artificial environment, designed for them to live comfortably beneath the lunar surface. The idea of a domed environment is very much like what is being proposed for a future lunar colony from Earth. In addition, there are some who believe that humans have been existing on the Moon in an artificial domed environment that is run by a secret space program for years. Domes on the Moon were also seen and commented on by the *Apollo 16* astronauts.

> **Duke**: We felt it under our feet. It's a soft spot. Firmer. Where we stand, I tell you one thing. If this place had air, it'd sure be beautiful. It's beautiful with or without air. The scenery up on top of Stone Mountain, you'd have to be there to see this to believe it—those domes are incredible!
>
> **MC**: O.K., could you take a look at that smoky area there and see what you can see on the face?
>
> **Duke**: Beyond the domes, the structure goes almost into the ravine that I described, and one goes to the top. In the northeast wall of the ravine, you can't see the delineation. To the northeast there are tunnels, to the north they are dipping east to about 30 degrees.

What were the astronauts witnessing? Were they talking about a structure with domes? In *Elder Gods of Antiquity,* (page 101), author M. Don Shorn writes, "Other anomalies include what appears to be 'domes' formed in discernible clusters, not in random patterns of natural distribution. Such domes are large, gently sloped structures, which are mostly located in or near craters." States contactee Alex Collier in his "Moon and Mars Lecture 1996," "According to the Andromedans there were primarily nine huge dome cities on the moon, in the back they housed up to 5 million extraterrestrials at one time. There was water, there was vegetation, there was everything they needed. These covered hundreds of miles." *Did someone beat us to the Moon and establish a domed city?*

Lunar Constructions Discovered

What appear to be intelligently made constructions were discovered in 44 different locations on the lunar surface. The following is a small list of the more well-known objects. They have been given names by dedicated lunar researchers that painstakingly sifted through thousands of NASA images. Two of the major contributors for bringing this information to the public's attention are Richard Hoagland and Mike Bara, who are experts on anomalous lunar structures.

An Ancient Complex

In their video *The Moon Pictures*, the Broadcast Team Alpha Show team reveal an interesting image taken by one of the Chinese space probes. The picture includes what seems to be an ancient building complex. According to the video, the image includes a "multilayered square cornered building. In the lower half it looks like a row of smaller buildings and to the right it looks like streets or smaller buildings."

An Ancient Spaceship

An object that appears to be a spacecraft that was discovered in NASA photographs taken during the *Apollo 15* mission. This spawned a hoax about an *Apollo 20* mission. According to the story of *Apollo 20*, three astronauts were sent on a secret mission to the Moon to locate what officials believed might be an ancient spaceship (located in the *Apollo 15* photograph). This alleged spacecraft is enormous and believed to have possibly been a mothership.

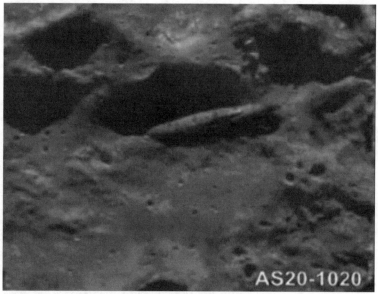

The photo of a spaceship on the Moon allegedly from *Apollo 15* used in the so-callled *Apollo 20* hoax (?).

The Bridge

What is believed to have been an intelligently made bridge on the Moon was discovered on July 29, 1953 by American astronomer John O'Neill, a former science editor for the *New York Herald Tribune*. The structure was twelve miles long and connected two mountain peaks that were located on the rim of the Sea of Crises (Mare Crisium). When O'Neill came forward with his discovery, he was ridiculed by associates. However, after British astronomer Hugh Percy Wilkins was called in by O'Neill to examine the region where the bridge had been spotted, Wilkins corroborated the sighting.

English astronomer Patrick Moore stated that the bridge had appeared out of nowhere, as this area of the Moon had been studied before, and there was no bridge there previously. This generated a great deal of discussion about the strange anomaly. After a few years, it suddenly vanished. The disappearance of the bridge led to speculation over what happened to it. Years later, researchers brought forward an interesting theory. Some hypothesized that the bridge was disassembled by extraterrestrials to keep their existence a secret from humans on Earth. Some believe that the extraterrestrials had somehow discovered that humans were observing it. Today, we have moved beyond whether the bridge was real. We are now at the point of trying to understand if Earth is being monitored by aliens, and just who put the bridge there, and why.

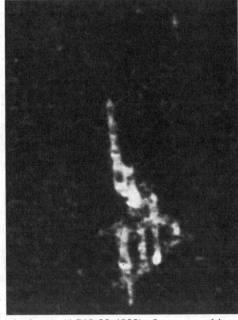

The Castle

In an image taken by the *Apollo 10* mission, there is an object that has been dubbed the castle. The castle is a bright, oddly shaped

An image (AS10-32-4822) of a strange object taken during the Apollo 10 mission. It has been dubbed The Castle.

structure that is suspended several miles above the lunar floor. The bottom has a row of what look like pillars, and there is a feature on top that resembles a turret. There have been speculations as to what this strange object is.

The Chalet

A structure located in the Tycho crater. An image of the item is featured in author Mike Bara's book *Ancient Aliens on the Moon* (page 186). In his book, Bara states that the object "looks somewhat like an A-frame building." Bara located the "chalet" in a photograph taken by the *Clementine* spacecraft.

The Condorcet Hotel

An anomalous construction found on the Moon by the *Apollo 17* astronauts. The Condorcet Hotel was mentioned in a conversation between the *Apollo 17* astronauts, while working on the lunar surface. The astronauts' puzzling conversation only enhanced the mystery of the *Apollo 17* mission, as they talked amongst themselves about something extraordinary that they were witnessing.

MC: Go ahead, Ron.

Evans: O.K., Robert, I guess the big thing I want to report from the back side is that I took another look at th—the—cloverleaf in Aitken with the binocs. And that southern dome (garble) to the east.

MC: We copy that, Ron. Is there any difference in the color of the dome and the Mare Aitken there?

Evans: Yes, there is... That Condor, Condorsey, or Condorecet or whatever you want to call it there. Condorcet Hotel is the one that has got the diamond shaped fill down in the uh—floor.

MC: Robert. Understand. Condorcet Hotel.

Evans: Condor. Condorset. Alpha. They've either caught a landslide on it or it's got a—and it doesn't look like (garble) in the other side of the wall in the northwest side.

MC: O.K., we copy that Northwest wall of Condorcet A.

Evans: The area is oval or elliptical in shape. Of course, the ellipse is toward the top.

In her book, *ET's Are on the Moon & Mars*, C.L. Turnage comments on this odd exchange between the astronauts stating, "Who is the mysterious 'they've' who have caught a landslide?" She alludes to the idea that the astronauts were inadvertently suggesting that the Condorcet Hotel was manufactured by unknown race.

The Geo-Dome

An anomalous construction located on the rim of the Tycho crater. In his book *Ancient Aliens on the Moon* (page 187), Mike Bara describes it as "roughly hexagonal in shape," and says it "seems to jut from the hillside as if it were the tip of a much larger subsurface object." The geo-dome is also described as having a roof, a wall, and windows.

Glass and Crystal Structures

There has been mention of what appears to be intelligently made glass or crystal-like constructions on the Moon. One of these structures was mentioned during the *Apollo 16* mission by astronaut Charles Duke, where he talks about seeing what resembles a beautiful, part glass, part crystal-like structure, located at the Descartes site. He states, "Right out there... the blue one I described from the *Lunar Module* window is colored, because it is glass coated. But underneath the glass it is crystalline...the same texture as the Genesis rock... dead on my mark." Some believe that the astronauts at that moment were looking at an artificial construction of a glass-like structure. In this case, we are speaking of what appears to have been a highly evolved construction process made from some sort of glass or glass-like material. Some craters appear to have a sort of glassy substance on the bottom as well. It has been speculated that some of the glassy areas seen in craters may be covering a secret city or town located beneath the Moon's surface.

Madler's Square

An anomalous geometric construction situated close to Mare Frigoris. It is named after German astronomer Johann Madler.

Madler's Square.

Because of its design and perfect angles, some believe it to be an artificial structure that dates back to ancient times. It has extremely high walls and is in an angular-shaped pattern that forms a square. In *Elder Gods of Antiquity* (page 101), author M. Don Schorn sites a quote by the famed British astronomer Edmund Neison from his book titled *The Moon* (1876). Neison states that Madler's Square is "a perfect square, enclosed by long, straight walls about 65 miles in length and one mile in breath, from 250 to 300 feet in height." This strange formation has been discussed in many books in the past about the Moon. Some question whether it may be from a civilization located there ages ago.

Obelisks

Several obelisks are located on Earth, and some have been located on the Moon. The reason for the ancients creating obelisks is unknown on Earth. The fact that some were discovered on the Moon is even more mysterious. Is there a connection? Could it be possible that they were created by the same beings ages ago that were traveling between Earth and the Moon? The most well-known obelisks on the Moon are those dubbed The Blair Cuspids. The Blair Cuspids are structures that appear to be ancient monuments. They are six formations that were captured in images taken by

A photo of the obelisks on the Moon called the Blair Cuspids.

NASA's Lunar Orbiter 2 in the Moon's Sea of Tranquility region.

They resemble the iconic Washington Monument in Washington, DC. They appear to have precise shapes and patterns and are in perfectly aligned geometric positions. They are named after anthropologist, William Blair who worked for the Boeing Institute of Biotechnology who examined the photographs taken by Lunar Orbiter 2. His comments about the cuspids appeared in a *Washington Post* article titled "Six Mysterious Statuesque Shadows Photographed on the Moon by Orbiter," on November 22, 1966. In the article Blair stated, "If the cuspids really were the result of some geophysical event, it would be natural to expect to see them distributed at random. As a result, the triangulation would be scalene or irregular, whereas those concerning the lunar object lead to a basilary system, with coordinate x,y,z to the right angle, six isosceles triangles and two axes consisting of three points each."

Given that there appears to be an ancient Earth-Moon connection, with the possibility that our ancient ancestors were visited by beings with aerial technology, it is no surprise that we have found objects resembling obelisks, pyramids, and other objects familiar to us. We can only imagine the purpose of these monuments. Those on the Moon may even have similar cultural significances as the ones on Earth. In an article titled, "Mysterious 'Monuments' on the Moon," *Argosy Magazine,* August 1970, Volume 371, Number 2, Writer and Paranormal Researcher Ivan T. Sanderson asked, "Is the origin of the obelisks on this Earth and those on the moon, the same? Could both be ancient markers originally erected by alien space travelers for guidance of later arrivals?"

The Paperclip

The paperclip is a very odd, extremely large construction discovered in a NASA photograph by Researcher and Author, Richard Hoagland. Hoagland named the object "the paperclip." The paperclip appears to be sitting on a pole and seems to be a part of some sort of assembly. Some lunar researchers believe it to be an unnatural object, while others are skeptical.

Pyramids

Pyramids are known to be mysterious, with their own hidden special meanings. On Earth, archaeologists, historians, and other experts are still trying to unlock the secrets to them. Imagine their surprise when researchers discovered several pyramidal shaped structures on the Moon. Some were photographed by NASA's Lunar Orbiter space probes during the 1960s. During the Apollo 17 mission, the crew took several photographs during their time on the surface. Amazingly, one such photo includes what seems to be a rather large pyramid. Another picture of a pyramid was taken by the Hubble telescope. In this image one can even see the Pyramid's shadow on the surface. In his book *Ancient Aliens on the Moon,* author Mike Bara writes about a

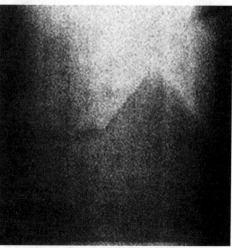

An image of a pyramid taken during the *Apollo 17* mission.

construction in the Tycho crater that is shaped like a pyramid. He describes it as "very bright and appears to have a 4-sided pyramidal structure." One theory for the pyramidal structures is that they are directly connected to highly advanced, intelligent beings that came into our solar system long ago and established colonies on the Moon, Earth, and Mars. Since pyramids are found all over the Earth, with the most famous being the great pyramids of Giza, the discovery of pyramids on the Moon poses several questions such as, "Is there any connection to the pyramids on the Moon and those on Earth? Were they created by the same builders?"

Some researchers of this strange coincidence believe that the connection lies with a prehistoric culture on Earth that had the ability of spaceflight. Interestingly, the pyramids of Giza, Egypt are believed by some to be connected to the cosmos and have a unique position where the three pyramids of Giza line up perfectly with the three stars of Orion's belt. Could this be the case with the pyramids on the Moon? Do the pyramids have an Orion connection? In the video titled *The Moon Pictures,* a NASA image with a pyramid is featured. In referencing the pyramid, two thought-provoking questions are asked, "Who are the pyramid builders?" and "You don't build pyramids in spacesuits, do you?"

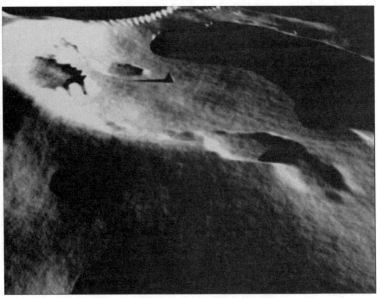

NASA photo of a pyramid on the Moon.

Roads

According to sources, what appear to be roadways have been located on the Moon. These mysterious roads are said to run through craters, hills, and valleys. One was measured at being approximately 62 miles long. Some have speculated that the road-like patterns show that someone with vehicles has been moving across the lunar surface.

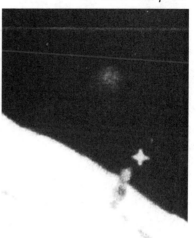

The Shard

The object known as the Shard is a large, mysterious erect edifice in the shape of a club. Located in

An anomalous object on the Moon. It has been dubbed the Shard.

the Moon's Ukert region, it is 1.5 miles high. It was famously named "the Shard" by Writer and Researcher Richard Hoagland.

The Sphinx

An enormous, sphinx-like figure was discovered in the Tycho crater. It is estimated to be approximately 230 feet (70 meters) in height and bizarrely resembles the sphinx of Egypt. Ancient Egypt is believed by some lunar researchers to have a mysterious cosmic connection. There are some who believe that the formation resembling a sphinx is evidence of an ancient extraterrestrial connection between the Moon and the Earth. Purportedly, the framework of the sphinx-like structure is organized in the same manner as the one in Egypt. It is estimated to be proportional in size, with the same lines and a similarly shaped head. One noted difference is that the Moon's sphinx is estimated to be even older than the one in Egypt!

There is the idea that the pyramids of Egypt align with stars. There is another idea that beings gave the ancient Egyptians and other cultures of the time information to build the pyramids and cross between dimensions; it is even said that some of the extraterrestrials lived among the ancient Egyptian people as rulers.

Could the same beings that are believed to have interacted with the ancient Egyptian society be connected to the galactic Earth-Moon history? Are these same beings responsible for introducing aerial technology to the Egyptians as seen on the ceiling beams in the temple of Seti I in Abydos? More importantly, where are the creators of this sphinx-like object on the Moon *now*?

Spaceports

The prominent Archimedes crater could be a primary area where a spaceport is located. There, large cylindrical-shaped objects and strange lights have been spotted. These objects have been estimated to be approximately twenty miles long and three miles wide. It is thought that they may be extraterrestrial spacecraft. In her book *ET's Are on the Moon & Mars,* C.L. Turnage writes, "A strange collection of gigantic, cylindrical objects can be seen resting on the floor of the crater Archimedes." Earth has been visited by UFOs that fit this description. Some believe that it is possible that the same objects seen in the skies around Earth, may be coming in from the Moon.

The Spike or Antenna.

In a NASA image there appears to be what resembles a spike rising up from the lunar floor. It is estimated to be three to four hundred feet tall, and is clearly artificially made. It has also been referred to as an antenna.

Structure with a D-shaped Wall.

In Rima Hadley (a sinuous rille) sits a mysterious construction surrounded by a D-shaped wall. It is in the higher part of Rima Hadley, near where *Apollo 15* sat down.

The Tower

The tower is an object photographed by NASA's Lunar Orbiter 3 which launched on February 5,

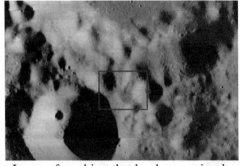

Image of an object that has been variously referred to as "the spike," and "the antenna."

1967. It is over a mile wide and stands approximately five miles above the lunar floor. It was dubbed the "tower" by Writer and Researcher Richard Hoagland. One wonders what a structure on the Moon like this could be used

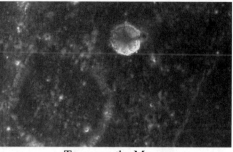

Tower on the Moon.

for. Is it a lookout point? A landmark? Are there extraterrestrials inside?

The Tower of Babel

A structure shaped like a tower was photographed on the Moon by the Soviet space probe Zond 3, which was launched in 1965. Zond 3 was to be sent towards Mars as a spacecraft test, and to explore interplanetary space. As it travelled around the far side of the Moon, it captured spectacular images that were sent back to Earth. One of the images was of great interest to scientists. It was an image of a formation that appeared to be artificial. It was estimated to be 3.5 miles in height. The construction is projected up and out of a crater. It was nicknamed, the Tower of Babel. The researchers were perplexed by this structure and found themselves unable to explain its origin or its meaning. The Tower of Babel can be seen in NASA photograph No. AS10-32-4856. There have been

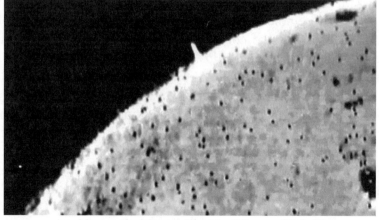

An object on the Moon dubbed the "Tower of Babel."

147

many speculations about this anomaly. Some ufologists believe that it may have a connection to an extraterrestrial base, while others maintain that it is an artifact from an ancient lunar city.

The Trapdoor

There is a precisely created, very large square located inside the Bancroft crater. It has been referred to as a "trapdoor." It is suggested that spacecraft can enter the door and there may possibly be a secret base located underneath. The Broadcast Team Alpha team feature it in their video titled *The Moon Pictures*. It can be located on Google Moon at 27degrees, 54, 52.71, North 6 degrees, 23, 23.8.

Water Tanks

Two oddly shaped structures can be located in the *Apollo 16* photo NASA No. 16-18918. One researcher described them as resembling water tanks. Elsewhere they are referred to as towers. Whatever these objects may be used for on the Moon, they can be seen easily in the image. They are said to be very large and reaching upward from the lunar surface. They appear to have a semicircular covering at the top.

The above objects were given names to match what they resemble to us, not necessarily for what they are. We simply do not know for certain what they are, nor what their functions may be. One of the most interesting features of an alleged extraterrestrial civilization is that the description of the Moon artifacts and ruins *sometimes* resemble the architecture from prominent cities of ancient Earth. Then of course, there are others that we may not even recognize as ancient ruins. In fact, Farouk El Baz, who was a leading geologist during the Apollo program is quoted in a 1974 article in *SAGA Magazine,* where he states, "We may be looking at artifacts from extraterrestrial visitors without recognizing them." As Dylan Clearfield, in his book, *Alien Threat from the Moon,* so aptly put it, "The identity of some of the objects is obvious, some require speculation and some—while clearly visible—defy explanation. They are there, but they simply escape definition."

Extraterrestrial Bases

Some believe that extraterrestrials have established a base on the far side of the Moon. According to the book *The Source: Journey Through the Unexplained* (page 125), by Researchers Art Bell and Brad Steiger, "From time to time, alleged Air Force veterans,

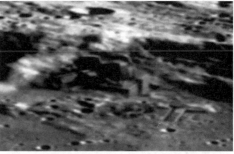

An image of a structure on the Moon taken by the Chinese Chang'e-2 Orbiter.

insisting upon complete anonymity, swear that the United States established a secret base on the Moon in 1970 and that they discovered an alien base on its dark side." This base has been nicknamed "Luna." It is believed to have facilities for staff as they engage in a mining operation. This mining operation is believed to have been going on for centuries, as they extract needed resources for their home planet or for existing on the Moon. It is also an area where extraterrestrial spacecraft are said to be parked. There have been sightings over the years of UFOs traveling to the far side of the Moon. Strange, unexplainable formations have been in pictures from the far side, thus sparking the theory. There are those that even propose that as well as small ships, there are large motherships located there. There is a story of a NASA employee who discovered a base in Apollo photographs in the 1960s. Supposedly, the pictures showed an extensive complex that had structures and constructions in a variety of shapes and sizes. Some were described as having geometric shapes, and there were towers as well as mushroom- and spherical-shaped buildings. The idea of cities beneath the surface of the Moon conjures up visions of large, beautiful, technologically advanced areas.

The Cosmic Highway

Could extraterrestrials be traveling through "shortcuts" in space to reach the Moon and possibly Earth? These shortcuts are otherwise known as stargates. Since it can take thousands of years to traverse the universe, this just may be the method that extraterrestrials are using to cover vast distances quickly. The astronauts of *Apollo*

17 took photographs of what some researchers believe could be a stargate on the Moon. NASA image AS17-151-23127, shows a mysterious, illuminated object on the Moon. In an article titled "Did Apollo 17 find a Stargate on the Moon?," Researcher and

An image of a structure shaped like wheels. It was taken by a Russian probe.

Author Michael Salla writes, "The object appears to be a space portal of some kind with an eerie blue glowing ring around a central darker portion. The object looks similar to the travel device depicted in the fictional film *Stargate*."

The idea of a stargate on the Moon is a theory that many will believe should remain in science fiction. However, from what we have learned about potential extraterrestrials, UFOs, Tic Tacs, space travel and more, we should remain open to the possibilities. In fact, there is a theory that the UFOs we see today are coming through stargates, which are also known as portals, gateways, and doorways. They have even been referred to as cosmic highways.

The Definitions website offers this description of a stargate: "A hypothetical device consisting of a traversable portal (typically a wormhole) that can send one to another location light years away nearly instantaneously." In other words, a stargate is essentially a portal to other worlds. Although these stargates are for the most part hypothetical, we are becoming more aware of the prospect of extraterrestrials possibly having the ability to move rapidly via shortcuts in space. Perhaps one of the reasons that we see UFOs on the Moon is due to them coming in from stargates before traveling to Earth or elsewhere in the solar system. This would again fall under the "stopover" scenario on the Moon. On Earth there are said to be areas that have portals as well. Iraq is thought to be one of them. Peru, the Himalayas, the Amazon, Antarctica, Siberia, and the Grand Canyon have also been named as places on Earth with possible stargates. There are ancient tales that tell of people entering and exiting Earth through "spiritual doorways"… stargates. Could the same be happening with our Moon and other

Puerta de Hayu Marca (Gate of the Gods). Peru.

areas of the galaxy?

Perhaps the most well-known portal on Earth is the one located in Peru. Some call it the "Peruvian stargate." The Peruvian stargate is a large construction that was found in the Hayu Marca mountains of Southern Peru. There it is referred to as the "Puerta de Hayu Marca" (Gate of the Gods). The door was carved out of a stone mountain ages ago. It measures approximately 23 feet high and 23 feet wide. There is a smaller recess in the center at the base measuring approximately two meters tall. According to legend, during ancient times men passed through this doorway by supernatural means to visit the "land of the gods." Purportedly, there was an invisible energy force that allowed a physical human to enter the stone energetically and travel to other worlds via mystical tunnels within. What is interesting is that there are accounts in recent years of people passing through this doorway.

Is this stargate a figment wild imagination? If real, is it something akin to the transporter in *Star Trek*? Could we have advanced technology here on Earth remaining from our distant past that allows us to travel great distances, and to other worlds? And most importantly as it applies here, could beings have been traveling back and forth into space and to the Moon via these superior transportation devices? It is thought that the gods travelled

151

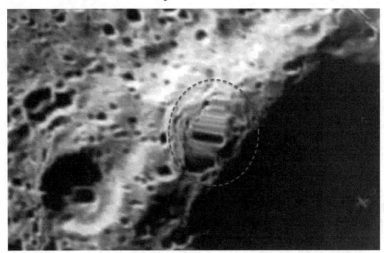

Odd sturture on the Moon. The opening is thought to be for ships coming
and going.

back and forth between the Moon and Earth at some point in time.

Could this have been one of the ways they transported? Even more amazing is that there have been rumors for several years now of a government having a classified space program that has access to portal technology. This means that someone, somewhere on Earth has invented a mechanism that allows human beings to travel to a different location, anywhere in the universe in a matter of what may be only minutes. If there is any truth to these stories, then perhaps this is exactly what happened on the Moon.

Humans could be there if we used this type of technology in the past. Beings far superior to us may also have this type of technology and may have been using it for eons, perfecting it as they go along. For those that are skeptical about such ideas, we should remember all is energy. Who is to say that we have not learned to manipulate our own energy to allow us to reach one point to the other in an instant? As I have pointed out before, we could have reached such technology ages ago... *and lost it.*

Industry on the Moon

Most cities have some sort of industry. However, I never thought I would be considering such an idea about the Moon! Oddly, there appears to be industry on the Moon. For example, it appears that someone is mining the Moon. As we have seen, the idea that the

Moon is being mined came from images from the *Clementine* and Apollo missions, where what "appears" to be mining equipment was found. Extraterrestrial contactee George Adamski mentioned that during his trip to the Moon, he had witnessed a barter system in place where supplies and provisions were exchanged for minerals mined from the Moon, with beings that had flown in on spacecraft. The supposition is that the Moon had been excavated for these materials. There is also the idea that extraterrestrials built installations on the Moon for the purposes of mining. According to ufologist Don Wilson, in his book *Our Mysterious Spaceship Moon*, what appeared to be excavating operations were spotted on the Moon by the Apollo astronauts. He writes, "The concentric hexahedral excavations and the tunnel entry at the terrace side can't be results of natural geological processes; instead, they look very much like open cast mines." Open cast mining is a surface mining technique in which large pieces of land are hollowed out to allow for extracting minerals. If this is happening on the Moon, then who is doing this excavating? Is this material being taken off world? Do they use it in their day-to-day existence? Or are they manufacturing something?

The famed remote viewer Ingo Swan (1933–2013), known as the best in his field during his time, remote viewed the Moon. He told of seeing extraterrestrial humanoids at work. Swann spoke of factories, humanoid beings, and robots creating a laser beam. Could these beings be mining the Moon as well? In his book *Aliens on the Moon* (page 20), author Bryan Kelly writes, "Industrial processes are in operation right now on the surface of the Moon. There are certain locations where heavy surface machinery may be regularly observed in action, some examples were actually photographed by Apollo crew members on a number of occasions." Such machines were seen in the vicinity of individual craters. At the Humboldt Crater, there appears to be an excavation operation. Reportedly, this operation can be seen on an *Apollo 15* photograph NASA No. 15 1512640. Helium 3 may be the source of interest for some of this activity.

When considering something as major as there being a metropolis on the Moon, it is important to look at all aspects of the information that we have available. The above is only a partial list of some of the very interesting evidence that tells us that the Moon is not only a place where we are being watched, but a place with a system. With our limited understanding of otherworldly beings, we can only look at areas we may have in common. It appears from looking at the above information, that a city-state may be one of those areas, be it on the surface, or be it a megastructure with an established habitat inside.

Even if we travelled to the Moon today and found no one resides there now, we can be certain from what we know that someone most certainly once did. Of that I am certain. However, there is still plenty of activity being seen on and in the skies of the Moon today. Additionally, there may be beings located on the Moon, as well as in areas above it that we have no knowledge of. It may also be possible that lunar residents have their own off planet megastructure, as there have been several occasions where people have mentioned seeing objects that extend for miles and then disappear. If these extraterrestrials have the technology to build a megastructure, to be self-sustaining, and to travel the cosmos, I have no doubt that their technology allows them to be "cloaked" to the point that we would not notice them in areas above the Moon. *Time will tell.*

The craters Aristillus, Aurolycus and Archimedes. The Aristillus crater seems to have a group of pyramids in the center of it.

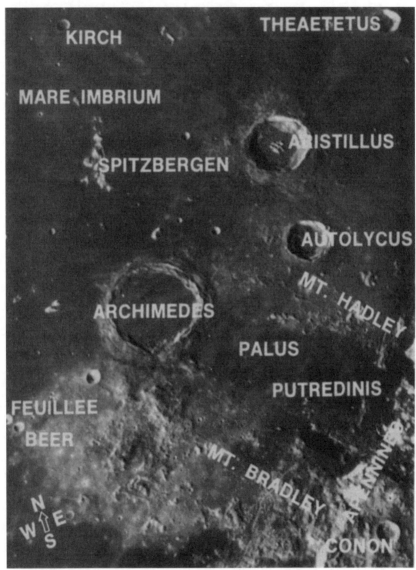

The craters Aristillus, Aurolycus and Archimedes are shown here with other features nearby.

Chapter Seven

Hark! Who Goes There?!

It is probable there may be inhabitants in this other World,
but of what kind they are is uncertain...
—John Wilkins, *Discovery of a New World in the Moon*

The Extraterrestrial Presence on the Moon

Many people today believe that extraterrestrials exist. Some
even believe that they are here on Earth, living among us. Others
maintain they are here on secret missions observing Earth and
humans, unseen. There are those that clam to have seen alien
spacecraft landing and otherworldly beings exiting from them.
Others maintain that benevolent, humanoid beings approached
them with messages for mankind. There are others that claim to
be hybrid children of an extraterrestrial and human union. The list
of stories about extraterrestrials is a long one! Several books have
been written about extraterrestrial and human encounters. In fact,
there are many books available today naming the various groups
of otherworldly beings that have visited Earth. Who are these
space beings? Where do they come from? Are the stories true?
Do any of them inhabit the Moon? If we return to the Moon and
discover that it really is inhabited, then who might we find there?
Are there beings today that are watching mankind from the Moon?

Because we have been taught that the Moon is uninhabitable, it
is difficult for people to believe that there could be life there, even
for those that believe that extraterrestrials exist. Most will say that
they believe that we are not alone in the universe given the vastness
of it, and that they understand that life is out there somewhere, but
they stop short when it comes to the Moon. Scientists tell us that
the Moon cannot harbor life. However, from examining the Moon,

it appears that *someone* is there. Some of the confusion stems from the contradictory information. The Moon is supposedly dead with no activity, yet astronomers and lunar researchers are stating the opposite.

Some maintain that the structures found in NASA images are the result of pareidolia. However, can we honestly say that every anomalous object discovered in images are examples of our seeing something that is not real? This is important to consider because even if one of the anomalies of the many I have already described is an intelligently created object, then that means we are not alone in the universe, and someone is on the Moon. It only takes one. Even if we were to find the Moon abandoned now, we still have enough evidence to at least contemplate that someone once existed there. The same applies for the Apollo missions to the Moon. Nearly every mission was followed by strange lights or a UFO. If just one of those accounts was not caused by a natural occurrence in space, then my friends, we are not alone in this universe.

It appears that there may have been extraterrestrials dwelling on the Moon for millions of years. If true, then it means that our search for extraterrestrials was always as close as the Moon, only 252,088 miles away at its most distant (225,623 at its closest). There are quite a few theories as to who may be there. It could be one of the groups listed here, or none of them at all. However, I feel that it is time to move past the question of "Is someone there?" and consider who they may be. Perhaps we can then figure out if they are friend or foe.

What is the Evidence that an Extraterrestrial Race is on the Moon?

Edgar Mitchell was the Lunar Module pilot for *Apollo 14*. He spent nine hours on the lunar surface. He once recalled an experience he had while working on the Moon. Mitchell commented, "I had to constantly turn my head around, because we felt we were not alone there. We had no choice but to pray." Later, after becoming a UFO proponent, Mitchell once commented, "I have no doubt that extraterrestrials could very well have populated or made structures on the far side of the moon." He should know. If we are to take the words of anyone seriously when it comes to this topic, then the Apollo astronauts are among those at the top of the list. However, not only the astronauts, but others that work in the field of space travel, astronomy and more that have put their reputations on the line when making such statements and presenting information. What is the evidence that someone is on the Moon? We have:

•Photographic evidence of structures and anomalous objects on the Moon.
•Eyewitnesses of UFOs following the NASA astronauts.
•Statements from Apollo astronauts of seeing UFOs, strange lights, and moving objects while in space and on the Moon, as well as receiving unexplained radio signals.
•Testimony from credible people who worked at NASA that came forward to tell of the Moon anomalies encountered by the astronauts and those found in photographic images.
•Accounts from astronomers and lunar researchers of their discoveries on the Moon.

The following are additional comments made by leading astronauts, astronomers, authors, NASA personnel, researchers, ufologists, and others regarding the Moon being inhabited. We will keep their thoughts in mind as we move forward into the information given in this book.

•"I figure it is logical to believe that spaceships might be using our moon for a base in their interplanetary travels." George Adamski, Ufologist

•"Our Moon is a potential indicator of a possible alien presence near the Earth at some time during the past 4 billion years." – Alexey Arkhipov, Radio Astronomer and Astrophysicist

•"At no time, when the astronauts were in space were they alone: there was a constant surveillance by UFOs." –Scott Carpenter, NASA Astronaut

•"Are we to conclude that these lunar craters have been frequented by extraterrestrial astronauts? ...The possibility cannot be rejected."

The Moon's Galactic History
–Robert Charroux, Author

•"Moments before Armstrong stepped down the ladder to set foot on the Moon two UFOs hovered overhead." –Maurice Chatelain, former NASA Communications Engineer

•"The Moon is firmly in the possession of these occupants. Evidence of Their presence is everywhere, on the surface, on the near side and the hidden side, in the craters, on the maria, and in the highlands." –George H. Leonard, Author, and Researcher

•"I have no doubt that extraterrestrials could very well have populated or made structures on the far side of the moon." –Edgar Dean Mitchell, Astronaut

•"Could it be that the Moon is a foreign country and someone else's property?" –Steve Omar, Author, and Researcher

•"Many phenomena observed on the lunar surface appear to have been devised by intelligent beings." –Ivan Terrance Sanderson, Author, and Researcher

•"Photos, both American and Soviet, reveal that seeming non-natural, artificially made structures do exist on the Moon!" –Don Wilson, Author, and Researcher

•"I advance the theory that our moon has been, and still probably is used as an advanced observation base, in regard to our Earth, by mysterious cosmic visitants connected with the flying saucer phenomena." –Harold T. Wilkins, Ufologist

•"It is probable there may be inhabitants in this other World, but of what kind they are is uncertain." John Wilkins, Astronomer

Who are the Moon's Inhabitants?
Major Donald Keyhoe (1897-1988) was an American Marine Corps Naval aviator. Additionally, Keyhoe authored several articles on aviation for several leading periodicals and became a recognized ufologist. He considered it vital that the U.S. government investigate UFOs. After research about the Moon,

he once commented, "All the evidence suggested not only the existence of a Moon base, but that operations by an intelligent race have already begun. If so, who could the creatures be? Were they from other planets or did they originate on the Moon?" If there are beings on the Moon who are they? There are several theories as to what race of beings may be residing on the Moon. These include the following.

Humans on the Moon

There is the possibility that if the Moon is inhabited, it is humans that occupy it. There are different scenarios behind this theory. The first stems from the legends of ancient times that speak of humans having knowledge of advanced aerial technology. Earlier we looked at the legends of the Hindus, Atlanteans and others that are believed to have had this knowledge. There is also evidence that humans may have reached the point of being an advanced civilization, to the point of traveling into space and as far as the Moon and perhaps even further. There is also evidence that Earth has had catastrophes in the past that nearly wiped out humanity, and those left had to start developing all over again. This may be where we are today. Is it possible that the beings on the Moon are humans from a time before who established settlements on the Moon and then were cut off from Earth, perhaps due to a worldwide disaster that prevented them from returning? Could they have become stuck in space? Perhaps this scenario could be applied to Mars and other planetary bodies within our solar system as well.

In a movie directed by actor George Clooney (in which he played the lead character) titled *The Midnight Sky,* a similar scenario was portrayed on screen. In the film, Clooney plays Augustine, a scientist working in the Arctic. Due to a worldwide catastrophe, Augustine must warn astronauts in space not to return to Earth. There may be a human colony on the Moon, with an artificial environment, that has been there for ages. If so, these humans would be very different from us. Are the beings rumored to have been encountered on the Moon by the *Apollo 11* mission from our ancient past? If so, this may the reason they were following the Apollo missions. Are they waiting for us to mature as a worldwide

161

culture before they make themselves known to us? They may have forbidden any contact with us until we are ready. This of course is all imaginative speculation.

Another scenario for humans being on the Moon is that of the governments from Earth placing secret colonies there. When we see the activity on the Moon, supposedly, this may be those that were sent there on secret missions to colonize the Moon. Some propose that it may be a German colony that was started in the 1940s. Allegedly, the Germans reverse engineered an extraterrestrial spaceship, which allowed them to travel into space. They are believed by some to be living on the far side of the Moon. Other governments are believed by some to have established a secret mode of travel to the Moon as well.

The theory is that they were able to establish settlements on the Moon. American aeronautical engineer Ben Rich is regarded as a brilliant scientist and is known as the "father of stealth." He was a former head engineer in charge of the Lockheed Skunk Works. He led the development of the F-117, the initial production stealth aircraft. He is a person that had inside information on our state-of-the-art science and technology, and possibly UFOs and extraterrestrials. Rich is quoted as stating in a speaking engagement, "We already have the means to travel among the stars."

Others have made the same claim. According to researcher and contactee Alex Collier, there are 35,000 human beings from earth that live on the Moon." This claim was made in a video of his titled "Moon and Mars Lecture, 1996." It has also been speculated that some missing people from Earth may have been abducted and taken to the Moon and other planetary bodies ages ago to populate those worlds. After looking at all these scenarios, it seems that there is a possibility (if even a remote one) that humans may already be on the Moon.

In their paper "Relationships with Inhabitants of Celestial Bodies," Robert Oppenheimer and Albert Einstein state, "Another possibility may exist, that a species of homo sapiens might have established themselves as an independent nation on another celestial body in our solar system and evolved culturally independently from ours." If Oppenheimer and Einstein were correct in their assessment, it would be interesting to consider what these people would be like in the areas of technology, culture, language, history, customs, and traditions. It would also be fascinating to learn what they know about Earth and humanity.

Time Travelers from the Future

In thbook *Who Built the Moon?* author Christopher Knight talks about his belief that the Moon was created to help life on Earth flourish, and that the creators were in all probability humans from the future. There are ancient artifacts and artwork that suggests that there were visitors from the past that may have been connected to Earth. A number of these relics appeared to have been created with technology far too advanced for that time. This shows that there may have been a civilization on Earth in the ancient past that had achieved advanced technology. This could have been the result of visiting extraterrestrials bringing knowledge of superior technology to humans from the future.

Besides the advanced aerial technology that was mentioned in Chapter Four (Ancient Aliens and the Earth Moon Connection), we should consider if this technology that was potentially given to humans could have taken people to the Moon and further into space, as there are images of people in spacesuits. Interestingly, the Zuni Native Americans use items in their rituals that imply that they had contact with either extraterrestrials or possibly time travelers. The ceremonial items consist of masks that resemble the astronaut helmets of today. The Hopi Native American tribe have kachinas (dolls) that symbolize extraterrestrials. There is an ancient tale out of Ravda, Bulgaria of blue star beings visiting Earth.

What is notable is that there are scientists who predict that humans born on another planet would eventually change appearance due to the foreign environment and may actually be blue in appearance. Could these beings have been time travelers from our future? Could they be humans from the past that adapted to a new environment and were visiting Earth? Could Earth play a part in the histories of extraterrestrials on other worlds beginning with the Moon? Was the Moon a jump-off port for an advanced technologically aware mankind of long ago, that has now seeded other planets? If any of these scenarios are true, then it just may be that our Moon is inhabited with humans. Additionally, the Moon may be a spaceport where humans from other planets are visiting and meeting before coming to Earth. With so many UFOs in Earth's skies these days, one wonders if there will be some sort of announcement made soon that will lead to disclosure. Could that be the reason for all of the activity on the Moon?

Humanoids

The Merriam-Webster dictionary defines a humanoid as something "having human form or characteristics." The SidmartinBio, Wide Base of Knowledge website takes it one step further stating that a humanoid is, "a non-human creature or being, or robot, with human form or characteristics." Humanoids are believed to be scattered throughout the galaxy and possibly seeded by the same beings. There are some who believe that humanoid extraterrestrials from other parts of the galaxy may be living among us observing our behavior here on Earth. It is also thought that humanoids may be on the Moon. There is a NASA image that shows what appears to be a humanoid on the Moon. There is a second image that can be seen on Google Earth showing what looks like someone traversing the Moon. Some argue that these are fibers on the camera lenses. Others believe they are aliens. However, the size of at least one of the figures is so tall that is has been dismissed by some as being a biological entity. In any event, the images are very odd and, as usual, fuel much speculation about aliens on the Moon.

Lunar Aboriginals

There may be lunar aboriginals. There is the idea that an advanced race of beings seeded orbs in the galaxy and that the Moon was one of them. Not only planets and planetoids were thought to have been seeded but other moons and asteroids as well. The Moon may have beings on it that have been there all along, developing on their own, just as humans do on Earth. They are believed to have been there from the beginning and are the original "Selenites." These may be the beings to whom the ruins found on the Moon belong, and that are responsible for the activity witnessed on the Moon for all of these centuries.

Extraterrestrials from Elsewhere

Another theory is that the original inhabitants are no longer there, having either died off or fled the Moon, and another race located the Moon and now occupy it. This new group may have been searching for a new home or are there to mine the Moon for its resources.

A Variety of Species/Races

It has been said that there are several different races existing on the Moon today. This could be a sophisticated populace with different races peacefully inhabiting the Moon. A twist on this idea is that this may also include humans.

The Fallen Angels

According to one theory, the Moon is where the "fallen angels" of biblical lore left Earth and escaped to during the flood. These fallen angels would have been alien in origin and not from the divine as has been taught. However, if this is true, then there may be a connection to the Moon and biblical stories.

Life on the Moon

We should remember that when scientists state that life cannot exist on the Moon, they are speaking of life as we understand it. This means that the Moon cannot generate or germinate life as we know it, in the state that it is in today. However, we have no idea what type of alien lifeforms may be out there. When it comes to extraterrestrials, we can only imagine what their characteristics would be, how they live, if they are similar to us, etc. There are several different types portrayed in the science fiction arena. It is not a stretch of the imagination to think that those from other worlds would have their own way of existing, and like earthlings may have bodies designed to live on their worlds. These beings could run the gamut in terms of their organic natures. Some of them (like us) may need special outerwear or a spacesuit of some type to help them live on a new world. Some may be naturally built to exist on other planets that are more suited to their makeup.

It could well be that, depending on their nature, they may be able to withstand the harsh elements of the Moon, or perhaps they discovered a way to do so. In fact, there could be extraterrestrials whose whole manner of existing would be different from ours. They may be highly developed enough to be able to create a false body, or a type of suit to aid them in existing on other worlds including the Moon. Some of these beings may also have dispatched AI's (artificial intelligence) to do their work. There could also be beings that are non-airbreathers. These would be beings that do not have the need or the constitution to breathe in air. There may also be beings that have no need for food. These beings may exist by ingesting some specific type of energy that acts in a way to sustain them. If that is the case, there may also be beings with no need for organs. They would originate from worlds where these are not needed to exist.

Such differences may make extraterrestrials immune to all sorts of dangers that would affect humans on the Moon. In commenting on this subject in his book *Somebody Else Is on the Moon,* George Leonard states, "'Radiation unscreened by an atmosphere would be deadly!' pontificates a scientist, forgetting that others may be inured to it, or that they may use protective measures, such as underground abodes." We can assume then that there may be several species of extraterrestrials that can withstand being on the lunar surface without protection.

There is no limit when it comes to imagining what an advanced alien race would be capable of. Perhaps they can construct a body that is part organic and part metal and modify it so that it is tailored to the signature of the individual that will inhabit the body. In their book *Extraterrestrial Civilizations on Earth*, (page 77), authors Cecelia Frances Page and Steve Omar, who researched the topic of extraterrestrials extensively, shared a similar thought, writing, "On many planets, moons and asteroids where there is little or no atmosphere, there are humanoids with no lungs, hard thick skin or who are made out of energy and look like colored light bodies." If we open our minds to the possibility that we are not alone in the universe, if we step out of our comfort zones and imagine the unimaginable, then we can consider all sorts of possibilities and scenarios when it comes to otherworldly beings, and how they may be able to exist on the Moon.

If this all sounds like something right out of science fiction, just read a comment from Ben Rich quoted in an article titled, 'Ben Rich, Lockheed Martin and UFOs,'" by the Gaia staff, December 3, 2019. Under the subheading "Ben Rich's Deathbed Confession," it quotes Rich as stating, "We have things out in the desert that are fifty years beyond what you can comprehend. If you have seen it on *Star Wars* or *Star Trek*, we've been there, done that, or decided it wasn't worth the effort. They have about forty-five hundred people at Lockheed Skunk Works. What have they been doing for the last 18 or 20 years? They're building something." This just tells us, that there are projects in the works today that we know nothing of when it comes to space, the Moon, the planets, and extraterrestrials.

The Seeding of Worlds

Could the galaxy have been seeded by superior beings from another universe? Could there be beings out there that have been around so long that they have advanced to the point that they have creator capabilities? Were the planets and other cosmic bodies seeded by beings such as this? Is this the reason that many contactees describe having seen otherworldly beings that are humanoid? This theory was brought to life on *Star Trek: The Next Generation* in an episode titled "The Chase" (Season 6, Episode 20). In this installment, four opposing alien races endeavor to resolve an ancient genetic mystery that involves them all. The chase ends when an artifact is discovered and provides the answers to their questions. The object contains a holographic image of a primordial being, left on a planet eons before for the various races of the galaxy to find. The holographic image that was recorded by a female scientist from an ancient humanoid race, explains their origins:

> You are wondering who we are. Why we have done this. How it has come that I stand before you, the image of a being from so long ago. Life evolved in my planet before all others in this part of the galaxy. We left our world, explored the stars and found non like ourselves. Our civilization thrived for ages. But what is the life of

one race compared to the vast stretches of cosmic time? We knew that one day we would be gone, that nothing of us would survive. So, we left you. Our scientists seeded the primordial oceans of many worlds where life was in its infancy. The sea codes directed your evolution towards a physical form resembling ours. This body you see before you which is of course shaped as yours is shaped; for you are the end result. The sea codes also contain this message which was scattered in fragments on many different worlds. It was our hope that you would have to come together in fellowship and companionship to hear this message. And if you can see and hear me, our hope has been fulfilled. You are a monument, not to our greatness but to our existence. That was our wish, that you too would know life and would keep alive our memory. There is something of us, in each of you. And so, something of you, in each other. Remember us.

If there are beings on the Moon, could they be a part of the group that may have seeded the solar system? Could these beings be a faction of a larger group that are from another world? The Moon could simply be one of many stations that are observation posts to keep watch over worlds that they seeded on various cosmic bodies. Although it may be difficult to image other planets in our solar system as desirable areas to live, you might be surprised to find that we do not know the history of each planet, nor the entire system as we may think. Probes are being sent out to explore the planets. We do not yet have all the data as to what happened in a universe that is estimated to be 13.8 billion years old, the Milky Way Galaxy at 13.2 billion years old and the solar system at 4.6 billion years old. (More is explained in Chapter 10, Into the Future.) All worlds do not have to be Edenic as Earth is, to be called home. It is theorized that this may have been the case with our galaxy. The planetary bodies in our galaxy may have been seeded by an advanced race of beings. They may be observing us from the Moon, and other races from other worlds from different vantage points such as other moons and asteroids, or even spaceships.

If we listen to the tales of the ancients about people that came

Extraterrestrials from *Star Trek*.

from the stars, they are described as being humanoid. Why is that? There are many who believe that if there is extraterrestrial life out there, they would look nothing like us. This appears not to be the case, at least not in our galaxy. There is a pattern of humanoid beings seen in our universe. It is possible that, just as in the fictional *Star Trek* universe, our galaxy is scattered with humanoids that came from the same race that seeded other worlds to keep their DNA, their line, and/or their species alive. It could also be that advanced beings simply wanted to give life. Humans just may be one race among many from the same creators. When I talk about creators, I am speaking of beings that are highly developed enough to create life. Therefore, I wonder if these hypothetical beings were to exist, would they or did they seed other worlds and terraform them as well? Is the entire universe by plan, from the largest planet to the smallest star?

There is a theory that Venus was once a planet with life as Earth is today. Mars is also believed to have held life at one time. There is speculation about the Moon in this regard as well. It is thought that the Moon was originally a planetoid, and for at least three billion years, it could have held life. It may have had an atmosphere and a substantial amount of water at one time. That is a theory. Could the Moon have been one of those worlds that was seeded long ago?

Perhaps that original colony is gone now due to a catastrophe of some sort. Could there be an Earth Moon connection in our DNA? According to a French contactee who goes by the name

Robert L, there is colonization going on by an "Intergalactic Federation." He claims that in 1969, he was taken to a secret area located in the Himalayas and became a part of a genetic experiment to seed human life on a planet in another galaxy. Robert L came forward in 2005 to tell his incredible story. Perhaps what Robert L experienced is an example of what happened out there in the universe ages ago. The Moon was possibly one of the places used in this seeding of our galaxy.

Possible Extraterrestrial Races on the Moon

In the book *The Only Planet of Choice, Essential Briefings from Deep Space* (page 90), authors Phyllis V. Schlemmer and Palden Jenkins write, "If you start to talk to some people about extraterrestrial intelligences, they may think you're crazy. It's a major taboo to lift them out of fantasyland and into the possible-reality zone….UFO sightings and encounters with ETs of various kinds do happen."

There are many reports of people meeting beings from other worlds. In recent years, people have come forward claiming to have had contact with them. Most do not believe them. I do, as there are things that I have seen that I cannot explain. I personally know very sane, educated, successful people who have brought me stories of the unexplained involving otherworldly experiences. There are even rumors of President Eisenhower meeting with extraterrestrials. With all the stories of extraterrestrial contact circulating today, people should realize that if even one is true, then there is other intelligent life in the universe, and we are being visited. In attempting to establish who may be on the Moon, we can at the very least look at some of the information that has been brought forward about various groups of extraterrestrials in the past few years. Could any of the following groups have a faction on the Moon?

The Andromedans

The Andromedans are a race of extraterrestrials that are from the Andromeda Galaxy, a barred spiral galaxy located in the constellation of Andromeda, located approximately 2.5 million light years away. The Andromedans are said to be highly intelligent.

They have also been described as energy beings, bipedal, and standing taller than humans. According to sources they are also known as the "Glass People" and the "Mirrored People." Some Andromedans are located in close proximity to the Moon. When I first learned about the Andromedans, I was struck by their being described as "glass or mirrored beings."

This description is close to the "extraterrestrial" that was allegedly seen by one of the Apollo astronauts during the *Apollo 11* moon landing. That being was described as ethereal, which in my opinion would be similar to a "glass" type of being. If the story is true, I wonder if it could have been one of the Andromedans.

The Anunnaki

Some claim that beings on the Moon may be the Anunnaki (a humanoid race of beings, that are believed to have come to Earth and were mistaken for gods). Researchers maintain that it may be one of their ancient cities that we are seeing by way of ruins on the Moon. The Anunnaki, whose name means "those of royal blood," are gods from ancient Sumerian religion. According to Sumerian lore, they came from the stars and created mankind to act as a slave race. The Sumerians were members of the world's first civilization that flourished circa 3500 to 2000 BC, in the southern half of Mesopotamia. They spoke an agglutinative, ergative language

A Sumerian cylinder seal with Annunaki gods depicted in a flying vehicle.

171

A Sumerian cylinder seal with Annunaki gods depicted, one with wings.

that was unrelated to any other known language. The writing system that they established (cuneiform) was later borrowed by the Babylonians and Assyrians.

According to Author Zecharia Sitchin, the Anunnaki initially arrived on Earth approximately 450,000 years ago from the planet Nibiru (also known as the Tenth Planet and Planet X). They came in search of gold, which they located and mined in Africa. The planet Nibiru itself takes an elliptical path, reaching Earth every 3,600 years. They have been referred to as "gods," "angels," "the watchers," "the Nephilim," and more recently "extraterrestrials." These deities of ancient Sumerian lore are thought to have played a major role in the ancient Earth Moon connection.

It has been hypothesized by some lunar researchers that the Moon is a spaceship brought into Earth's orbit by the Anunnaki. Some even believe that the Moon is the planet Nibiru itself, or at the very least a spaceship created by the Nibiruins and sent across the universe to Earth (perhaps for the resources, as it appears that they knew about them in advance). For clarification, Nibiru is thought to hold different races of beings, the Anunnaki being just one of them. The gods were believed to travel back and forth between Earth and the Moon.

In fact, according to ancient Sumerian lore, the Anunnaki created Adapa (Adam) the first human. This is thought to have possibly occurred on the Moon, due to Adam historically being known as a prophet of the Moon. In some cases, during warring on Earth, they were said to have escaped to the Moon until the warring subsided. According to lore, after a conflict among the Anunnaki elite over the future of mankind, the Anunnaki left Earth and the Moon, and returned to their home, the planet Nibiru. Some researchers maintain that there are still Anunnaki on the Moon, possibly living inside the Moon's interior and observing humanity's development. These beings are also thought to continue to travel back and forth between the Earth and the Moon.

In her book *ET's Are on the Moon & Mars* (page 7), Researcher, and author C.L. Turnage writes about the Anunnaki, stating that they "were actually present during the time our astronauts made their historic Lunar landings." Perhaps if this story is true, and there are still Anunnaki on the Moon, that could be one reason why humans may have been "warned off the Moon," as rumored about the *Apollo 11* mission. According to lore, there was an argument between the leaders of the Anunnaki. Some wanted humans to have free will, to live their own lives and develop accordingly. Eventually, humans were released, and supposedly this is the reason that humans are free on Earth today. Even though humans are free, they may not be allowed by the Anunnaki on the Moon. Perhaps that was the reasoning behind the warning.

Suddenly, after a millennium of time, the Anunnaki and their place in our history is becoming clearer. Could there be a reason for this information coming out? Could this be a part of the disclosure of who we are, where we come from, and if we are alone in the universe? Some question if the Anunnaki are still collecting resources from Earth (for example gold), storing them on the Moon, and then sending them off to Nibiru. Could that be the explanation for so many UFO sightings around the world? Also, could this be the reason behind the "fleets" of ships that we see leaving the Moon in videos and photographs?

The Arcturians

The Arcturians are described as an advanced race of beings from the fifth dimension. They are from the star Arcturus, located in the constellation, Bootes of the Milky Way Galaxy, some 37 light years from us. According to the article "The Arcturians" by Clifford Stone, on the BurlingtonUFOCenter Website (Spanish Version), *"The Book of Knowledge—The Keys of Enoch,* describes it [Arcturus] as the mid-way programming center used by the physical brotherhoods in this universe to govern the many rounds of experiments with physicals at this end of the galaxy."

The Arcturians that are located within Earth's vicinity are said to be on a gigantic mothership. They are a benevolent race here to aid and assist humans as we ascend into the fifth dimension. However, they do not interfere with our free will. Their spaceships are said to be highly sophisticated. The Arcturians are also acting as protectors of Earth by warding off negative ETs who would otherwise overtake earth with their superior starships.

The Arcturians have transmitted a special message to humans through the form of a book. They channeled this information through Dr. Norma Milanovich, who wrote the book *We, The Arcturians*. They have large, dark almond shaped eyes. They are described as being smaller than humans, around three to four feet tall, slim with skin of a greenish hue. Their looks are very similar to each other's which omits much of the competition we face here on Earth. They have three fingers and are telepathic. Ordinarily they live to be between three hundred and four hundred years of age. Sickness is unknown on their home world and their life is ended when their contract (their reason for coming into existence) is completed. In an article from the popular website anomalien. com titled "The Arcturians – The Most Evolved Alien Species in

Is there a Reptilian menace on the Moon?

174

Our Galaxy and Earth's Wardens" it states, "The Arcturians have built bases on our planet, particularly in secluded mountainous regions, but also on the surface of the Moon. From these strategic spots they carry important operations where they revitalize Earth's energy points that have gone inactive." It seems that the Arcturians are among the extraterrestrials that we can refer to as watchers and protectors.

The Reptilians

The Reptilians are a well-known race of malevolent aliens that some believe created the Moon eons ago. According to the Zulu people of South Africa, the Moon was brought here by two extraterrestrials that were reptilian. This then may be the most plausible group to look at when considering who may be on the Moon. They are

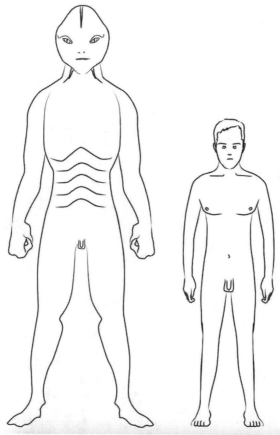

A comparison between a Reptilian and a human.

described as six to eight feet tall, bipedal, with green, and sometimes brown scaly skin. They have gold eyes resembling those of a cat. They are thought to exist on the Moon (some believe they are living underground). The Reptilians supposedly depend on Earth for food and water for their survival. Some ufologists and researchers claim that resources from Earth are being extracted by these beings on a regular basis, and that they do not want to be discovered. Thus, it is believed that there is a cover-up when it comes to these beings. There have been popular shows made about Reptilians masquerading as humans such as "V" (i.e., Visitors) (1983). In the show, the Visitors attempt to take over Earth. They end up having to fight a resistance. Images of reptilian beings in all forms are found throughout our history, even in the Bible. Some contend that the famous *Apollo 11* account of the astronauts seeing ships sitting on the rim of the craters may have described craft that belonged to the Reptilians.

The Zetans (or Zetas)

The Zetans are an extraterrestrial species from the star system Zeta Reticuli. They are better known as the greys. They have been linked to most of the alien abduction experiences reported around the world. As a result, we have intricate descriptions of these beings. They are described as having large black eyes with no pupils, large hairless heads, with just a slit for a mouth. Their ears are described as a small lump on each side of the head. For clothing, they wear what appears to be tight-fitting silver or light

A Zetan.

blue jumpsuits with a high neck. The Zetans have no lungs and therefore do not require an atmosphere to exist and no digestive system. They are said to be emotionless creatures, lacking in spirituality. They are advanced in technology and science and are superior in intellect.

One of the most famous abduction cases is that of Betty and Barney Hill, a couple that were abducted in the 1960s by the Zetans, and whose case has been well documented. Betty Hill was shown a star map by the Zetans that abducted her and her husband, to show her where they were from. Hill later drew a map of Zeta Reticuli when relaying her abduction experience. There is a theory that the reason the Zetans abduct humans is to harvest human DNA, which they are believed to be using in the repairing of their genetic material.

Some ufologists believe that the Zetans inhabit the far side of the Moon, where they are believed to have a secret base from which they observe humans. These are thought by some to be the beings that warned the astronauts to stay off the Moon. In his article, "U.F.O. and Reported Extraterrestrial On... Moon and Mars," researcher and author Steve Omar writes a great deal about the Zetans. He postulates their reasons for being on the Moon saying, "They would die in our Earth atmosphere without a space suit." He continues with, "Our Earth atmosphere is poisonous to Zetans." Purportedly, the Zetans have several cities belowground, as well as a military establishment, a science installation, mining stations and a spaceport located on the far side of the Moon.

Watchers and Sentries

An article appeared in *Life* magazine (January 17, 1969) titled, "Our Journey to the Moon," written by the astronauts Frank Borman, Jim Lovell, Bill Anders. In it Frank Borman writes, "The view of the Earth from the Moon fascinated me—a small disk, 240,000 miles away. It was hard to think that that little thing held so many problems, so many frustrations. Raging nationalistic interests, famines, wars, pestilence don't show from that distance. I'm convinced that some wayward stranger in a spacecraft, coming from some other part of the heavens, could look at earth and never know that it was inhabited at all. But the same wayward stranger would certainly know instinctively that if the earth were inhabited, then the destinies of all who lived on it must inevitably be interwoven and joined. We are one hunk of ground, water, air, clouds, floating around in space. From out there it really is 'one world.'"

It appears that the view of Earth from outer space is a unique one. It is understood that all on Earth are one. The Earth is special, and there may be more going on with it than we are aware of. Our world still has a lot of unanswered questions about it. It is still a mystery, and the Moon may be connected. As a metaphysical researcher and writer, I have learned that the Earth is a very important planet. It isn't just some random planet in our solar system, where life accidentally developed. It, and all life on it, is considered unique and special by those on other worlds and even in the various dimensions.

This may be why there are so many tales of extraterrestrials coming to Earth in the past, and why there are so many accounts of UFOs and contactee experiences today. In the case of the contactees, there are various beings that visit, but always with the same message: that mankind should be careful to not destroy ourselves and other life on this planet. They say that our actions could have a rippling effect in the galaxy, most especially in the case of an atomic war. The Watchers is a name that I created for this unnamed group of beings that may be on the Moon, attempting to protect us from blowing ourselves up, and also from nefarious alien beings, wandering asteroids that could hit Earth, and other dangers. This group may not belong to any of the other groups named here as possible lunar inhabitant candidates.

They may, however, be playing a vital role in our lives, and they just may be doing it from the Moon. There may be beings out there that are assigned to the Moon to watch over Earth, acting as sentries. The same beings may have had a hand in the Moon being where it is, or not. In any case, there appear to be beings in outer space that are helping planet Earth, and this could be them. At least there may be a faction of them on the Moon. Who they are we do not know. This group of beings, I presume, may be on the Moon for the purpose of watching and protecting Earth but are to remain hidden until we are ready for open contact. Presumably, they are watching and waiting for the day that they will let us know of their existence and our place in the universe. This group probably considers us to be a young species, and are observing our scientific, technological, and even spiritual progress.

There have been tales of asteroids having near-Earth misses that could have done a lot of damage to the planet. There are some who believe that the reason Earth is not struck by these dangerous objects more often is due to the help of extraterrestrials. Some believe that Earth is under the protection of extraterrestrials that are in Earth's vicinity to assist us. Some reportedly have witnessed meteors exploding in the skies and witnessed a UFO nearby just after a meteor was blown to pieces. Since the Moon is the closest planetary body to Earth, we can only wonder if it has lunar inhabitants that may have a hand in destroying these dangerous cosmic entities before they hit our planet. In some cases, it is believed that there are extraterrestrials assisting Earth by pushing meteors and asteroids away from us. Some ufologists believe that one of the groups may be on the Moon and that a part of their reconnaissance mission is to be a sentry for those living on Earth. It is said that our galactic neighbors are waiting for us to advance technologically before we can move into the universal community. They are waiting for us to advance enough to protect ourselves from threats to Earth and all life on it. Sentries for Earth are believed to be placed strategically in our vicinity, including on the Moon. There is even speculation that NASA is a part of the plan and knows of the extraterrestrials' mission to protect us. A part of the reason it is thought that Moon inhabitants keep themselves hidden is because they are waiting for our advancement. It is thought that when the astronauts went up, they simply turned off the lights to remain undiscovered. There have been tales of extraterrestrials on and near the Moon seeking to assist humans should something go wrong. One of these stories comes from the *Apollo 13* mission, when the astronauts are believed to have been helped by extraterrestrials that were possibly aware of our space exploration endeavors. There was also the *Apollo 17* mission, where it is speculated that extraterrestrials aided the astronauts when they ran into a problem returning home.

The crew were evidently placed under orders not to mention

that they had received assistance from extraterrestrials to help them return to Earth. With all the talk of the *Apollo 11* mission astronauts being warned off the Moon, one former NASA employee, Maurice Chatelain, a former chief of NASA Communications Systems, wrote in his book *Our Cosmic Ancestors* that it appeared that the extraterrestrial spacecraft that were reportedly witnessed by the *Apollo 11* crew, appeared ready to assist the astronauts if something had gone wrong. With all the negative talk about extraterrestrials, it appears that we have cosmic friends and that quite possibly, some of them may be operating from the Moon. The Moon just may act as a point to hold the sentries assigned to our world. Life on Earth may not have been a random event as we were taught. There may have been beings that had a hand in much of the life that we see here, or even had a hand in terraforming this world. Advanced extraterrestrials may have a vested interest in Earth and want to protect it. The Moon would be the perfect vantage point from which to keep watch over Earth. Earth just may be the "child" of a group of beings that are observing us from afar, to see how we are faring.

Is there a Directive in Place?

Although it appears that there has been a cover-up regarding what has been discovered on the Moon, there is a possibility that it is not NASA or the world governments that have the intent of hiding things from the masses. It could be the lunar inhabitants themselves. Be they human or extraterrestrials or both, it could be them that are behind the silencing of information. While it is evident that there is someone on the Moon, it looks as if they are concealing themselves from us.

The non-disclosure of the existence of extraterrestrials may not be a matter of the governments of Earth not educating the public about them. The problem of non-disclosure may lie with the extraterrestrials themselves. It could well be that the extraterrestrials do not want us to know about them *just yet*. In an interview with Israel's *Yediot Aharonot* newspaper, former Israeli space security chief Haim Eshed, who authored the book *The Universe Beyond the Horizon, Conversations with Professor Haim Eshed,* states, "The Unidentified Flying Objects have asked not to publish that they are here, humanity is not ready yet." He also states that the extraterrestrials are seeking to comprehend "the fabric of the universe." Of what he has named the "Galactic Federation" (they are purportedly an alliance of space-traveling civilizations in the Milky Way Galaxy) Eshed stated, "They have been waiting for humanity to develop and reach a stage where we will understand, in general, what space and spaceships are."

To his credit, Eshed is a professor and former general. He oversaw Israel's space security program for almost 30 years. He is also a three-time recipient of Israel's Security Award. Evidently, when it came to a conversation about humans knowing of their existence, the extraterrestrials did not feel that it was time to disclose this information to humanity. How would they know this if they were not watching? Could it be then that disclosure is being held up for our benefit? They may see something in us that tells them that we need more time to process the life changing information that we are not alone in the universe.

This may well be the reason for us not having the proverbial "first contact" experience yet. Could the governments have made a deal with extraterrestrials not to tell the public, for our own benefit, thinking that we are not ready? For those of us who feel that we are ready to face the truth about there being other intelligent life in the universe, and our potential future with them, then perhaps we must slow down and wait for the world to catch up. Haim Eshed mentioned the "Galactic Federation." The name "Galactic Federation" has been around for years. For those following the news of outer space, potential extraterrestrials, and disclosure, you may recognize the name. Many, as I was, may have been shocked to hear the name in the mainstream news. As per usual, there has been no follow-up discussions as to who they are. *Will the world ever be ready?*

According to sources, otherworldly beings are educating us slowly about our future in space and are incorporating the idea of exploring the universe and making contact with other beings out there through the media and other means as we go about our daily lives. It is a program that appears to be quiet, easy, and subtle. This is especially true since the early 1900s. This cover-up, if coming from extraterrestrials, may involve potential extraterrestrials that are living or working on the Moon. It may also explain why the Moon appears to be a satellite from where we are watched. This may also be the reason that there is such confusion, and so many theories about what is going on with the Moon.

If there is an agreement in place between the governments on Earth and extraterrestrials not to disclose their existence, that may well be the reason for a lack of what some are calling substantial evidence that someone is on the Moon. It may answer the question as to why we do not see them openly. It may also be the reason that the astronauts were allegedly warned off the Moon. It may have had nothing to do with hostility, as some have asserted, against humans, but simply that there is a contract in place that tells others not to reveal themselves to us.

This may also have been the reason that none of the astronauts were directly approached during their visits to the Moon. It could simply be due to some sort of "directive" not to reveal themselves to us (if possible). Of course, if you have "menacing ships around," as purportedly announced by one of the astronauts, then perhaps that was a difficult thing to hide! (But I digress.) There have been rumors for years that some of the governments of Earth have been visited by advanced extraterrestrials. As stated above, there are even said to be agreements in place between them. This of course is hearsay and there is no physical evidence. However, if this were indeed true then consider the ramifications if this information got out. Would the people of this world be accepting of the news that there are extraterrestrials visiting, and that they have had communication with government officials? You and I may be accepting, but are *all* people ready?

Below is a document from June 1947 that was allegedly composed by Robert Oppenheimer and Albert Einstein about extraterrestrials and UFOs. It is titled "Relationships with Inhabitants of Celestial Bodies." The document was perceived as a guideline as to how to approach beings from another world. That of course includes the Moon. In 1947 the sky really was the limit and future space travel was on the horizon.

Relationships with Inhabitants of Celestial Bodies (June 1947)
By Dr. J. Robert Oppenheimer and Professor Albert Einstein

Relationships with extraterrestrial men, presents no basically new problem from the standpoint of international law; but the possibility of confronting intelligent beings that do not belong to the human race would bring up problems whose solution it is difficult to conceive.

In principle, there is no difficulty in accepting the possibility of coming to an understanding with them, and of establishing all kinds of relationships. The difficulty lies in trying to establish the principles on which these relationships should be based.

In the first place, it would be necessary to establish communication with them through some language or other, and afterwards, as a first condition for all intelligence, that they should have a psychology similar to that of men. At any rate, international law should make place for a new law on a different basis, and it might be called "Law Among Planetary Peoples." Obviously, the idea of revolutionizing international law to the point where it would be capable of coping with new situations would compel us to make a change in its structure, a change so basic that it would no longer be international law, that is to say, as it is conceived today, but something altogether different, so that it could no longer bear the same name.

If these intelligent beings were in possession of a more or less culture, and a more or less perfect political organization, they would have an absolute right to be recognized as independent and sovereign peoples, we would have to come to an agreement with them to establish the legal regulations upon which future relationships should be based, and it would be necessary to accept many of their principles.

Finally, if they should reject all peaceful cooperation and become an imminent threat to the earth, we would have the right to legitimate defense, but only insofar as would be necessary to annul this danger.

*****

Another possibility may exist, that a species of homo sapiens might have established themselves as an independent nation on

183

another celestial body in our solar system and evolved culturally independently from ours. Obviously, this possibility depends on many circumstances, whose conditions cannot yet be foreseen. However, we can make a study of the basis on which such a thing might have occurred.

In the first place, living conditions on these bodies, say the moon, or the planet Mars, would have to be such as to permit a stable, and to a certain extent, independent life, from an economic standpoint. Much has been speculated about the possibilities for life existing outside of our atmosphere and beyond, always hypothetically, and there are those who go so far as to give formulas for the creation of an artificial atmosphere on the moon, which undoubtedly have a certain scientific foundation, and which may one day come to light. Let's assume that magnesium silicates on the moon may exist and contain up to 13 per cent water. Using energy and machines brought to the moon, perhaps from a space station, the rooks could be broken up, pulverized, and then baked to drive off the water of crystallization. This could be collected and then decomposed into hydrogen and oxygen, using an electric current or the short wave radiation of the sun. The oxygen could be used for breathing purposes; the hydrogen might be used as a fuel.

In any case, if no existence is possible on celestial bodies except for enterprises for the exploration of their natural riches, with a continuous interchange of the men who work on them, unable to establish themselves there indefinitely and be able to live isolated life, independence will never take place.

*****

Now we come to the problem of determining what to do if the inhabitants of celestial bodies, or extraterrestrial biological entities (EBE) desire to settle here.

1. If they are politically organized and possess a certain culture similar to our own, they may be recognized as a independent people. They could consider what degree of development would be required on earth for colonizing.

2. If they consider our culture to be devoid of political unity,

they would have the right to colonize. Of course, this colonization cannot be conducted on classic lines.

A superior form of colonizing will have to be conceived, that could be a kind of tutelage, possibly through the tacit approval of the United Nations. But would the United Nations legally have the right of allowing such tutelage over us in such a fashion?

a. Although the United Nations is an international organization, there is no doubt that it would have no right of tutelage, since its domain does not extend beyond relationships between its members. It would have the right to intervene only if the relationships of a member nation with a celestial body affected another member nation with an extraterrestrial people is beyond the domain of the United Nations. But if these relationships entailed a conflict with another member nation, the United Nations would have the right to intervene.

b. If the United Nations were a supra-national organization, it would have competency to deal with all problems related to extraterrestrial peoples. Of course, even though it is merely an international organization, it could have this competence if its member states would be willing to recognize it.

*****

It is difficult to predict what the attitude of international law will be with regard to the occupation by celestial peoples of certain locations on our planet, but the only thing that can be foreseen is that there will be a profound change in traditional concepts.

We cannot exclude the possibility that a race of extraterrestrial people more advanced technologically and economically may take upon itself the right to occupy another celestial body. How, then, would this occupation come about?

1. The idea of exploitation by one celestial state would be rejected, they may think it would be advisable to grant it to all others capable of reaching another celestial body. But this would

be to maintain a situation of privilege for these states.

2. The division of a celestial body into zones and the distribution of them among other celestial states. This would present the problem of distribution. Moreover, other celestial states would he deprived of the possibility of owning an area, or if they were granted one it would involve complicated operations.

3. Indivisible co-sovereignty, giving each celestial state the right to make whatever use is most convenient to its interests, independently of the others. This would create a situation of anarchy, as the strongest one would win out in the end.

4. A moral entity? The most feasible solution it seem would be this one, submit an agreement providing for the peaceful absorption of a celestial race(s) in such a manner that our culture would remain intact with guarantees that their presence not be revealed.

Actually, we do not believe it necessary to go that far. It would merely be a matter of internationalizing celestial peoples and creating an international treaty instrument preventing exploitation of all nations belonging to the United Nations.

*****

Occupation by states here on earth, which has lost all interest for international law, since there were no more *res nullius* territories, is beginning to regain all its importance in cosmic international law.

Occupation consists in the appropriation by a state of *res nullius*.

Until the last century, occupation was the normal means of acquiring sovereignty over territories, when explorations made possible the discovery of new regions, either inhabited or in an elementary state of civilization.

The imperialist expansion of the states came to an end with the end of regions capable of being occupied, which have now been drained from the earth and exist only in interplanetary space where the celestial states present new problems.

*****

Res nullius is something that belongs to nobody such as the moon. In international law a celestial body is not subject to the sovereignty of any state is considered *res nullius*. If it could be established that a celestial body within our solar system such as our moon was, or is occupied by another celestial race, there could be no claim of *res nullius* by any state on earth (if that state should decide in the future to send explorers to lay claim to it). It would exist as *res communis,* that is that all celestial states have the same rights over it.

*****

And now to the final question of whether the presence of celestial *astroplanes* in our atmosphere is a direct result of our testing atomic weapons?

The presence of unidentified space craft flying in our atmosphere (and possibly maintaining orbits about our planet) is now, however, accepted as *defacto* by our military.

On every question of whether the United States will continue testing of fission bombs and develop fusion devices (hydrogen bombs), or reach an agreement to disarm and the exclusion of weapons that are too destructive, with the exception of chemical warfare, on which, by some miracle we cannot explain, an agreement has been reached, the lamentations of philosophers, the efforts of politicians, and the conferences of diplomats have been doomed to failure and have accomplished nothing.

The use of the atomic bomb combined with space vehicles poses a threat on a scale which makes it absolutely necessary to come to an agreement in this area. With the appearance of unidentified space vehicles (opinions are sharply divided as to their origin) over the skies of Europe and the United States has sustained an ineradicable fear, an anxiety about security, that is driving the great powers to make an effort to find a solution to the threat.

Military strategists foresee the use of space craft with nuclear warheads as the ultimate weapon of war. Even the deployment of artificial satellites for intelligence gathering and target selection is not far off. The military importance of space vehicles, satellites as well as rockets is indisputable, since they project war from the horizontal plane to the vertical plane in its fullest sense. Attack no longer comes from an exclusive direction, nor from a determined country, but from the sky, with the practical impossibility of determining who the aggressor is, how to intercept the attack, or how to effect immediate reprisals. These problems are compounded further by identification. How does the air defense radar operator identify, or more precisely, classify his target?

At present, we can breath a little easier knowing that slow moving bombers are the mode of delivery of atomic bombs that can be detected by long-range early warning radar. But what do we do in, lets say ten years from now? When artificial satellites and missiles find their place in space, we must consider the potential threat that unidentified spacecraft pose. One must consider the fact that misidentification of these space craft for a intercontinental missile in a re-entry phase of flight could lead to accidental nuclear war with horrible consequences.

Lastly, we should consider the possibility that our atmospheric tests of late could have influenced the arrival of celestial scrutiny. They could have been curious or even alarmed by such activity (and rightly so, for the Russians would make every effort to observe and record such tests).

In conclusion, it is our professional opinion based on submitted data that this situation is extremely perilous, and measures must be taken to rectify a very serious problem are very apparent,

Respectfully,

/s/

Dr. J. Robert Oppenheimer
Director of the Institute of Advanced Studies
Princeton, New Jersey

/s/

Professor Albert Einstein
Princeton, New Jersey

Chapter Eight

Symbols, Codes, Clues...

"There does not seem to be any height or elevation nearby from which the stones could have been rolled and scattered into this geometric form."
—Dr. A. Bruenko, Russian Engineer

During the first mission to the Moon in 1969, the *Apollo 11* astronauts left items on the Moon for future visitors. One of those items was a plaque that had written on it, "Here men from the planet Earth first set foot upon the Moon, July 1969, A.D. We came in peace for all mankind." It was signed by Neil Armstrong, Michael Collins, Edwin Aldrin, Jr., and President Richard Nixon. They also left a small silicon disc, approximately the size of a half dollar that carried messages of hope and congratulations from world leaders of 73 countries to the United States and the *Apollo 11* team. The disc was placed inside an aluminum container, then a pouch and was left on the Moon by the *Apollo 11* astronauts at the Sea of Tranquility along with other articles. At the top of the disk there is an inscription that states: "Goodwill messages from around the world brought to the Moon by the astronauts of *Apollo 11*."

A second inscription can be found along the edge of the disc. It states: "From Planet Earth –July 1969." From the wording of these items, it appears that the astronauts may have left them for extraterrestrials that may one day visit the Moon. They may have felt that this would give them an introduction to Earth and humans. If they were discovered, we can only imagine what extraterrestrials would have thought. Would they even be able interpret them? If so, would they attempt to respond? We should ask ourselves, would an alien race of beings on the Moon do the same? Might there be

The plaque left on the Moon by the *Apollo 11* astronauts.

Silicon disc left on the Moon by the *Apollo 11* astronauts.

symbols, codes and clues on the Moon left there for us to find, to let us know that they are there, or have been there? Not just us, but any group of alien beings that might visit the Moon, or simply be passing through.

Perhaps extraterrestrials left certain objects for us to find, that same goal in mind. This could be the case with some of the monuments, obelisks, and other objects discovered there. Is the Moon being used as a communication board to communicate with humans and other extraterrestrial races? In some of the mysterious occurrences on the Moon, lunar researchers have come across possible messages in various forms from otherworldly beings. There is no way, of course, to know if these are from past or current lunar inhabitants, or those that are simply stopping through, as our own astronauts did. Might there be symbols, codes, and clues to be found on the Moon, left there for lunar visitors of the future by other races? It could be that lunar people may be attempting to let us know of their existence or trying to alert us to some danger via signals.

Mysterious flashing lights have been observed on the lunar surface. Reports of these intermittent lights date back to the 1800s. On July 4, 1832, British Astronomer Thomas Webb observed a series of flashing dots and dashes that were similar to Morse code. On October 20, 1824, European Moon observers saw alternating blinking lights, occurring continuously throughout the night, close to the Aristarchus crater. In 1873, after carrying out a comprehensive investigation of the flashing lights, the Royal Society of Britain announced that the lights were being emitted by extraterrestrials attempting to signal Earth. Ken Mattingly, who was the Command Module Pilot during the *Apollo 16* Moon mission, described seeing unexplained flashes of light on the far side of the Moon during two consecutive orbits. Some astronomers speculate that the flashing is a type of code. Some have even suggested they are akin to a form of Morse Code, with the idea being that a message in Morse Code would be received much faster than radio messages between the Moon and Earth. Could it be that there is someone on the Moon attempting to signal? Might it even be a distress signal? We can only speculate on what someone may be trying to tell us. Were they trying to let us know of their existence? Or could it have been some sort of warning?

An example of flashing lights on the Moon can be seen in a video made by Ruben Ariza from Armenia, Quindio, Colombia. It was featured on the UFO Sightings channel titled "Strange Flashing Lights Sighted on the Moon" (September 2, 2021). Some propose that there may be a conflict or even a war happening on the Moon, and explosions may be the source of the blinking lights. British Scientist and Astronomer Cecil Maxwell Cade (1918–1985), who was once a member of the prestigious Royal Astronomical Society, wrote about seeing mysterious lights on the Moon. In his book *Other Worlds Than Ours: The Problem of Life in the Universe* (1966), Cade spoke of seeing irregular lights on the Moon, and their potential meaning. He stated, "Star-like lights, which could not have been due to the Sun's rays illuminating the tops of high mountains, have been the subject of many hundreds of observations; in fact, up to April 1871, no fewer than 1600 observations had been made for the crater Plato alone. Nor were these lights always single, or in small irregular groups, for many of the reports refer to 'geometrical arrangements.'" Cade considered what the lights could mean, asking, "Were these attempts at signaling by the inhabitants of, or visitors to the Moon?" *Good question.*

Aside from the lights, there also appear to be messages in some formations. An example of that can be found in Mare Imbrium. In Mare Imbrium, there are formations that appear to be a type of code. Author and Researcher Dylan Clearfield states in his article for *Exemplore*, titled "Ancient Cities Found on the Moon" (June 26, 2020), "Particularly alarming are codes formed out of the very landscape in Mare Imbrium by an unknown source. These are huge stone formations that have changed in shape over the years and since there isn't any weather on the moon the only cause of these changes would have to be by mechanical means. The meaning of the messages is not yet deciphered but they seem to be aimed earthward due to their size and position on the lunar terrain. Are they messages to secret extraterrestrial operatives on earth?"

Charles Fort (1874-1932) was a popular Author and Researcher whose specialty was anomalous phenomena. Fort spoke of moon inhabitants attempting to signal Earth. In his work titled

New Lands, Fort spoke of fluctuating geometric shapes being seen in the Linne crater. He believed this was a way that lunar inhabitants might be attempting to message us, stating, "Astronomers have thought of trying to communicate with Mars or the moon by means of great geometric constructions placed conspicuously, but there is nothing so attractive to attention as change, and a formation that would appear and disappear would enhance the geometric with the

Old print of the Man in the Moon.

dynamic. That the units of the changing compositions that covered Linne were the lunarians themselves—that Linne was terraced—hosts of the inhabitants of the moon standing upon ridges of their Cheops of the Serene Sea, some of them dressed in white and standing in a border, and some of them dressed in black, centering upon the apex, or the dark material of the apex left clear for the contrast, all of them unified in a hope of conveying an impression of the geometric, as the product of design, and distinguishable from the topographic, to the shining god [Earth] that makes the stars of their heaven marginal."

Writers Christopher Knight and Alan Butler, in their book titled *Who Built the Moon,* spoke of how the Moon was built as a message for intelligent beings on Earth stating, "It is designed to be meaningful only to intelligent creatures living on the Earth's surface."

If we accept the hypothesis that the Moon is not a natural satellite, but instead was intelligently created by advanced extraterrestrials, then we must consider whether the Moon's creators may have left us a message in some form. These extraterrestrial scientists may have left messages for us to find as we progressed over time. Even more interesting is that they may have left us clues to our origins.

This scenario is similar to that in the popular sci-fi fantasy *2001 A Space Odyssey,* where the story involves the locating of an ancient relic left on the Moon by advanced extraterrestrials millions of years before it was found. It was placed there to track mankind's development, and to signal when they had progressed to space travel. Even more amazing is that the movie reflects the theory held today that there may be advanced extraterrestrials monitoring mankind's progress into space from the Moon.

Clues That NASA Astronauts Were Not Alone

There have been bizarre incidents in which the astronauts heard voices in unrecognizable languages coming over the radio, along with strange noises. Are these clues that there was someone there that was attempting to communicate with the astronauts? Did the Apollo astronauts consider this when encountering such odd occurrences during their missions? At one point during the *Apollo 8* mission, the astronauts heard strange, garbled sounds coming from the radio. There was no explanation as to what the perplexing noises meant or where they originated. Additionally, there was a peculiar high-frequency radio noise that the astronauts described as being "intolerable." It seems almost as if someone were trying to get their attention. The astronauts, who are well-trained for various situations, were at a loss as to what was happening.

Could there have been alien beings on the Moon that were somehow trying to send the astronauts a message? Could the "intolerable" noise they heard have been a signal for them to stay away from the Moon? Could the pre-Moon-landing space missions have been warned to stay away from the beginning, and we didn't get the message because we didn't understand the clues? After all, we have no idea what a message from an alien would sound like. We do not have a "universal translator," as in *Star Trek.* A similar incident occurred with the *Apollo 12* mission when again, the crew and ground control heard inexplicable, baffling sounds that have been referred to as an unidentified language. The source could not be located at that time either.

Astronaut Al Worden (1932–2020) received a message in an unknown language while in the *Apollo 15* Command Module. According to the story, Worden heard a mysterious transmission

Astronaut Al Worden during the *Apollo 15* mission.

over the radio in what is believed to have been a possible alien language. He first heard breathing, then a whistling sound, and then someone speaking. The words were repeated continuously. The mysterious transmission was recorded, and Worden sent it to NASA. The words were, "Mara Rabbi Allardi Dini Endavour Esa Couns Alim." The message is said to have been censored and halted from being released. Oddly, the message was later broadcast in the media in France, from a possible leak. Professional linguists investigated the transmission but were unable to interpret it.

In an earlier incident, years before, NASA's *Mercury 9* may have received a signal from extraterrestrials attempting to contact them with an unknown message as well. During *Mercury 9,* which launched on May 15, 1963, a mysterious voice in a language that was not known on Earth was picked up on a special radio frequency by astronaut Gordon Cooper while passing over Hawaii. Later, the recording was examined, and it was determined that no known earthly language matched what was on the tape. In addition, as he crossed over Australia, Cooper saw a huge green UFO that

Apollo 10 astronauts Stafford and Cernan preparing for the mission.

was also picked up by tracking stations. No official word on what the UFO was. UFO proponents believe that beings from another world are especially interested in man's progression into space, and during that period, their potential future trips to the Moon. It is thought that what Cooper picked up on the radio, as well as the UFO, may have had a connection to the hypothesis that ETs from the Moon were watching. Perhaps they were sending clues to their existence and trying to communicate. Perhaps all of these incidents were connected to beings on the Moon, attempting to make a connection. In the end, we should ask, "was someone trying to tell us something from the Moon?"

Nearly four decades after the *Apollo 10* mission, recordings were found that indicated something strange had happened as *Apollo 10* traveled around the far side of the Moon. The *Apollo 10* astronauts had heard strange noises when orbiting the Moon. The men were perplexed at what they were hearing and referred to these peculiar sounds as "space music." After the astronauts returned from their trip, the tapes from onboard the spacecraft were transcribed and filed away inside of the NASA archives until 2008. These tapes are said to contain weird "music" that came from the radio. The noises were strange and out of place. The astronauts did not know what to make of it. The only thing they knew for sure was that no one would believe them, and that they should keep what happened to themselves. The following is dialogue taken from the transcripts:

LMP (Lunar Module Pilot Cernan: That music even sounds outer–spacey, doesn't it? You hear that? That whistling sound?
CDR (Commander Tom Stafford): Yes.
LMP: Whoooooo.
CMP (Command Module Pilot John Young): Did you hear that whistling sound, too?
LMP: Yes, Sounds like—you know, outer space type music.
CMP: I wonder what it is.

CDR: What the hell was that gurgling noise?
LMP: I don't know. But I'll tell you, that eerie music is what's bothering me.
CMP: God damn, I heard it, too.
LMP: You know that was funny. That's just like something from outer space, really. Who's going to believe it?
CMP: Nobody. Shall we tell them about it?
LMP: I don't know. We ought to think about it some.

Is it possible that extraterrestrials on the Moon were attempting to contact the *Apollo 10* astronauts? Perhaps they were trying to alert them to their presence. One wonders if the lunar inhabitants had been trying to communicate with the astronauts since the beginning of the space program.

There have been instances when unknown writings were purportedly seen on the Moon. In his book *Alien Threat from the Moon*, Dylan Clearfield tells of locating what seems to be writing on the lunar surface in Mare Imbrium. Clearfield writes, "It appears to be some form of writing, but in an unknown script. It almost looks like the word Greene." He continues, asking, "How would it be possible for a design like this to have formed on a lunar surface which has no active processes of erosion?" In his book *Secrets of our Spaceship Moon,* Ifologist Don Wilson writes of a strange

The Eratosthenes crater.

Moon signaling occurrence out of Japan. "One of the strangest of all such lunar reports comes out of Japan, where *Mainichi*, one of Japan's largest newspapers, reported the unusual discovery of Dr. Kenzahuro Toyoda of Menjii University, who, while studying the Moon through a telescope on the night of September 29, 1958, spotted what appeared to be huge black letters, so pronounced they were easily discernible. The letters seemed to form two words: PYAX and JWA. No one to this day knows what these letters on the Moon mean or can give an explanation of the experience."

Additionally, near the Moon's Littrow crater there is writing that appears to be the Greek letter gamma. The *Astronomical Register*, Volume 20, reports, "on the floor of the crater Littrow are seven spots in the shape of the Greek capital Gamma." There is a belief that Earth and the Moon were connected during ancient times. It has been speculated that the gamma was possibly put there by an ancient people that once existed on the Moon, possibly arriving there from Earth. There is also the case of the letter "X." The letter X has mysteriously been spotted in the Eratosthenes crater. UFO proponents maintain that this is a sign that there may have been a base in the area at one time. Additionally, in NASA

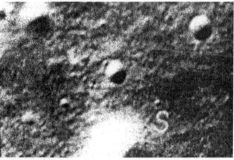

NASA image with a large mysterious letter S.

image No. 14-80-10439, there is an enormous feature in the shape of the letter "S." It unmistakably looks as if it were made by artificial means. Perhaps it is being used as a symbol that has an unknown message.

Radio signals have been picked up in several areas of the Moon in the past. They were detected and reported in the Moon's vicinity in the years 1927, 1928, 1934, and 1935. In 1935, two scientists by the names of Stormer and Van der Pol picked up radio signals both on the Moon and in the area around it. In 1958, astronomers in America, Britain and the Soviet Union located a UFO that was heading in the direction of the Moon. It was reportedly travelling at a speed of 25,000 mph. Radio signals were emanating from it. In 1956 several observatories picked up a signal that sounded as if it could have been a coded message that was being emitted from the Moon. The code was unable to be deciphered. One wonders if these were extraterrestrials signaling each other, or perhaps were signals meant for Earth.

In the possible signaling from lunar inhabitants, there are several strange occurrences to report. Although it may seem like a stretch to many that these could be possible signs from the Moon either to us or other aliens, they are still worth looking at. Of course, if there are lunar inhabitants, these events could simply be them going about their daily business. On July 6, 1954, Astronomer and former Curator at the Darling Observatory in Duluth, Minnesota, Frank Halstead (1883–1967) along with several others, witnessed an odd, straight black line inside the Piccolomini crater (located in the southeastern region of the Moon). This weird anomaly was not something that had been seen in the area before. The odd line was also confirmed by other astronomers. There is also the tale of an even more weird sighting, which is that of the "triangular protuberances."

In the book *Strange Universe*, by American Physicist and Author William R. Corliss, there is an astonishing account of the Moon having large triangular protuberances attached to it. The story was retold in *Reader's Digest, Mysteries of the Unexplained*. According to the account, on the night of July 3, 1882, a group of people who lived in Lebanon, Connecticut, USA, observed a peculiar occurrence on the Moon. It was explained that two, dark, obelisk-shaped protuberances materialized on the Moon's upper extremity. It is said that with the weird knobs, the Moon resembled a horned owl. A few minutes after they appeared, the protuberances slowly vanished, with the one on the southeast side dissipating first. Minutes later, two other obelisk-shaped protuberances appeared. These were located on the lower portion of the Moon. The notches slowly moved in the direction of each other, next to the rim. It looked as though they were eliminating almost a quarter of the Moon's surface as they were moving, until they finally came together. At that point, the Moon suddenly changed back to its natural appearance. According to *Mysteries of the Unexplained*, "When the notches were nearing each other the part of the moon

Some of the mysterious lines at Nazca, Peru including a hummingbird image.

seen between them was in the form of a dove's tail."

This incident certainly defies explanation. Not only that, but it was large enough to be seen by the naked eye. This could well have been an elaborate sign from lunar inhabitants. Could there be more signs from lunar inhabitants that we have missed? Mysterious lines were found on the Moon that have been compared to the Nazca lines of Peru. Some even refer to them as the Nazca lines of the Moon. In Peru, the lines were discovered in 1939 by Paul Kosok, an American water irrigation specialist, as he flew over the area. They are one of Earth's greatest mysteries. On Earth the lines are formed in geoglyphs, geometric shapes, animals, and straight lines. The Nazca lines on Earth are huge with some images measuring six hundred feet across. Our focus here is the straight lines. There are hundreds of straight lines, some over 30 miles long. They are believed to have been made between 400 and 650 AD.

On the Moon there are no depictions of animals, however, the straight lines of Nazca, compared with those on the Moon are eerily similar. Although they were originally believed to have been created by the Nazca people, some wonder if there were other people that created the lines, or if the Nazca people had some association with advanced otherworldly beings, as there is no logical explanation for these extremely large, artistic lines. The mystery therefore leaves researchers to speculate that they may have an ancient alien connection and may have been created by the same beings on the Moon. It opens a whole host of questions.

Given the history of spaceflight among the "gods" in ancient times, could it have been possible that these were the same beings that were traversing the cosmos and traveling back and forth between the Earth and Moon as stated in some of the Sanskrit writings of the past? Could these beings have been using the lines as guidance or a landing port on both the Earth and the Moon? Is it possible that they were trying to send each other messages within the lines and in the image of the animals? Some believe the Nazca lines were signs for aliens and perhaps may have been landing sites. Could these alleged beings have been located on the Moon in the first place? Could these lines on Earth and the Moon have been used for secret signs between the two planetary bodies? The world may never know the answers to these questions.

There is a strange story that comes from an astronomer in Germany. He was sent three different videos from people in three different countries including Japan, Germany, and Peru. Each video showed what appeared to be a large, black bird flying across the face of the Moon. In each of the videos, the image was the same. Since we know it is an impossibility for there to be birds on the Moon, we should consider if the birds are symbols for us to see? If so, what might they represent and what is the message?

Interestingly, Author C.L. Turnage, in her book *ET's Are on the Moon and Mars*, writes about what appeared to be five symbols seen in the Archimedes crater. Turnage believed that these symbols were meant to be seen from space. One of these symbols is a bird. She likens the bird to an eagle and ponders if there is a connection between the symbol of an eagle on the Moon, and the astronauts piloting the first lunar module which was named the Eagle. They declared the now iconic phrase, "The Eagle has landed." One can only wonder if these two strange occurrences with birds are connected, and if lunar inhabitants may be familiar with our symbols and imagery and are attempting to alert us to their presence with known symbolism. This also offers other food for thought, that is, are there Moon operatives on Earth gathering information about us and reporting back to the Moon?

Constructions in the shape of a cross are among symbols that have been seen on the Moon. They are so odd that they almost appear as if they were created to be noticed; it has been speculated that this would be a way that extraterrestrials could be attempting to alert Earth to their presence. One formation in the Kepler crater purportedly has an odd unnatural formation near it that is in the shape of a cross. It is huge, measuring four miles long and half a mile high. Some wonder if it was placed there by extraterrestrials long ago. In the book *Elder Gods of Antiquity* (page 102), Author M. Don Schorn writes, "Author Peter Kolosimo reported that photographs of strange structures in the shape of a cross, were taken by Astronomer R.E. Curtis and published in the *Harvard University Review*. They were deemed to be too purposefully arranged to be merely natural formations."

The Eratosthenes crater also has a cross-shaped anomaly. The *Astronomical Register,* Volume 20, states that there is "a geometric object shaped like a cross, in the lunar crater Eratosthenes." On the evening of November 26, 1956 an astronomer by the name of Robert E. Curtis, from Alamogordo, New Mexico, inadvertently took a picture of something highly unusual on the Moon. Curtis was studying the Moon, all the while taking pictures. Later, after developing the film, he was stunned to see what appeared to be a large, glowing, perfect white cross. Each branch of the cross was estimated to be several miles long. Each point of it was centered at the exact location to form a cross. Scientists tried to explain it away as the edges of mountains crisscrossing at the precise angles to give the illusion of a cross.

Author Frank Edwards commented on the subject in his book *Strange World* (page 275), saying, "Unfortunately for that explanation, it is physically impossible for mountain ridges to cross each other at right angles." One astronomer told of seeing objects that he referred to as a white cross and a square, in close proximity of the shaded area of the Moon as recently as 2018. Is there any connection between these formations? Do they have a meaning of some sort that is directed toward Earth? I find it interesting that the researchers here find them to be unnatural and have made notations about them, which lends credibility to them being of an unnatural origin. Unfortunately, we currently have no interpretation of their meaning; however, time may tell.

Portrait of Johann Carl Friedrich Gauss by Christian Albrecht Jensen.

Signals from Earth

Since the late 19th and early 20th centuries, there have been people with the idea of contacting extraterrestrials that may exist on the Moon and Mars. There were a few that brought forward ideas as to how to accomplish such a lofty goal, with some even making attempts at it. One such person was Johann Carl Friedrich Gauss (1777–1855), an accomplished German astronomer, mathematician, and physicist. Gauss believed that both the Moon and Mars were inhabited. As a result, he aimed his attempts at communication towards them. He created an elaborate signaling system to contact these beings. It is known as "Gauss's Pythagorean right triangle proposal." It revolved around the assumption that the Pythagorean theorem is a

A portrait of Nikola Tesla from 1919.

mathematical concept that should be ubiquitously understood by all advanced alien life forms. The proposal was to create a sufficiently large visual representation of the Pythagorean theorem on the surface of the Earth that it could be easily observed from space. This would be accomplished by planting enormously large fields of wheat in the middle of a pine forest. The shape of the wheat fields would form the typical geometrical representation of the Pythagorean theorem, where three lines are positioned in such a way that their intersections form a right triangle. He believed that this would convey a strong message that there is an advanced civilization on earth with a solid grasp of mathematics (geometry) and agriculture.

Serbian American Inventor and Mechanical Engineer Nikola Tesla (1856–1943) experimented in transmitting radio signals to the Moon to communicate with extraterrestrials. Tesla was overt about what he was doing, stating publicly that he was sending signals to the Moon. In the end, Tesla, fully believed that he had received a response from lunar inhabitants. Austrian Astronomer Joseph Littrow (1781-1840) also sought to alert possible extraterrestrials on the Moon to our existence on Earth. He created an extravagant system of communication that involved digging out trenches in various geometric shapes that included circles, squares, and triangles in the Sahara Desert. He had them filled with water, topped them with kerosene and ignited them, watching the flames fill the nighttime sky. It was his hope that extraterrestrials would see them and be alerted to our presence here on Earth.

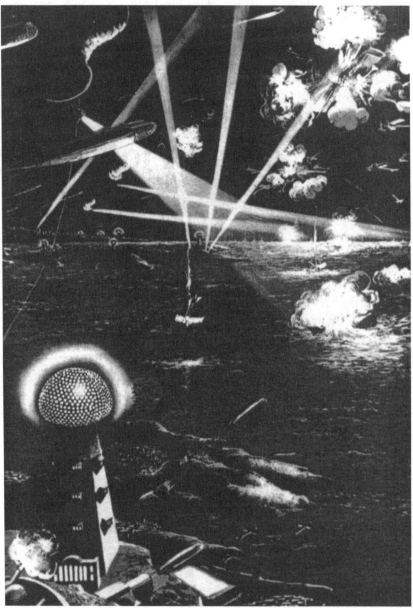

An illustration of Nikola Tesla's anti-gravity airships and broadcasting tower.

Chapter Nine

Where No Man Had Gone Before

"The Earth is the cradle of mankind,
but mankind cannot stay in the cradle forever."
—Konstantine Tsiolkovsky

On July 29, 1958, the National Aeronautics and Space Administration (NASA) was established. The original reason for the creation of NASA was to react to the Soviet Union's sending their satellite "Sputnik" into space, which occurred in October of 1957. In his now famous Moon speech, President Kennedy stated, "We choose to go to the moon. We choose to go to the moon in this decade and do the other things, not because they are easy, but because they are hard, because that goal will serve to organize and measure the best of our energies and skills, because that challenge is one that we are willing to accept, one we are unwilling to postpone, and one which we intend to win, and the others, too."

The Apollo Program was the third spaceflight program developed by NASA. The program was named after Apollo, the ancient Greek god of the Sun. In art, Apollo is seen traversing the skies in his chariot. This depiction was apropos for the missions that would literally launch men into space to explore the universe. The program was given its name by Abe Silverstein, an American engineer who was the former Director of Space Flight Development at NASA. The Apollo program ran from 1969-

1972, eventually placing twelve people on the lunar surface in six different missions! Twelve astronauts walked on the lunar surface, and six drove the Lunar Rovers. Three of the astronauts were fortunate enough to journey to the Moon twice. However, none of them landed on the Moon more than once.

The United States and the Soviet Union had competed to be the first country to place men on the lunar surface. President Kennedy wanted to accomplish this within the time frame of the 1960s, famously stating, "I believe that this nation should commit itself, before this decade is out, of landing a man on the moon and returning him safely to the earth …no single space project in this period will be more impressive to mankind or more important for the long range exploration of space." *"The long range exploration of space."* We had big plans. There were plans to have multiple missions to the Moon, learn as much as we could, and place human colonies in space, starting with the Moon, then Mars and perhaps beyond. Those first years of the Apollo program were an exciting, celebratory time. Space was surely our *final frontier*.

On July 20, 1969, Neil Armstrong (1930–2012) of *Apollo 11* made history when he took the first step on the Moon stating, "That's one small step for man, one giant leap for mankind." There was a global audience for this momentous event. With great anticipation, millions of people watched from around the globe. When the Lunar Module, the *Eagle* (aka LM-5) set down on the lunar surface, Armstrong famously declared, "The Eagle has landed," to let the world know that they had landed safely, and that their journey to the Moon was a success.

When the men landed thunderous applause erupted throughout mission control. Later, Neil Armstrong and Buzz Aldrin explored areas of the Sea of Tranquility, took photographs, performed scientific tests, and collected lunar samples from the Moon's surface. They even planted a United States flag to signify accomplishing the extraordinary feat of landing men on the Moon. It was a happy day! Mankind had achieved something astonishing. It was an enormous cerebral and technological accomplishment. Humans, in just a short period on a developmental and evolutionary timetable, had managed to launch themselves into the cosmos and land on another world.

Little did they know that the journeys into space would be filled with mystery and intrigue, and that eventually, the lunar missions would be cut short. Despite the fanfare, time, money, and effort (not to mention sacrifices, as sadly we lost the astronauts of *Apollo 1* in the beginning of this endeavor), we stopped going to the Moon. After *Apollo 17*, all planned missions to the Moon came to a halt. The reasons given were cost and loss of interest by the American public. *Loss of interest?* Were the missions not bigger than simply the public's interest? Did these missions not have a more existential cause?

It was not only a pipe dream, or a space race to the Moon. There was a hope of finding another planet to call home as we expand as a species and explore the universe. The cancelling of the missions to the Moon sparked a long line of conspiracy theories as people attempted to understand why the missions stopped. Some believed that something happened out there. Slowly, bits and pieces of a very strange story began to manifest. There were tales from the astronauts, and other people who worked at NASA during that period, that implied that there were unexpected events during the Moon missions. As a result, people began to question the motives behind NASA cancelling the last three lunar missions. What was it that could prompt the space agency to cancel something as important as space travel? Was it really cost and loss of interest?

There were rumors that the astronauts had encountered extraterrestrial ships while in space. Incredibly, it was also said that they had located evidence of an ancient civilization on the Moon. Today, with more information that has come out over the years, there are those who believe that that is the real reason why we stopped sending people to the Moon. What's even more amazing about this scenario is that the astronauts only visited less than one percent of the lunar surface. There may be more to discover there. If true, there are those who believe that whoever is there may be unfriendly, and not open to mankind visiting the Moon. This is because there is a story that the astronauts were "warned off the Moon." Is there any truth to story? Or is it just a rumor that is part of a conspiracy theory? If we listen to the strange tales that have come out of the Apollo program missions, and there are many, we can bet that it started with *Apollo 11*. That seems to have been the beginning of the end. What happened up there to make an entire nation change its course? Are the rumors of the astronauts encountering otherworldly ships and beings out there true? Did NASA learn something that we were not ready to hear fifty years ago? *Are we ready now?*

Apollo Missions

There are many mysterious stories involving the astronauts during the lunar missions. There are accounts of the astronauts being followed by unexplained lights and UFOs, seeing intelligently made structures and objects on the lunar surface, hearing unrecognizable language coming from over the radio systems and other perplexing anomalous events. There is even a rumor that one of the astronauts saw an extraterrestrial standing near one of the lunar modules. The bulk of this information comes from the astronauts, Apollo transcripts, and people that worked in the Apollo program. In his book *We Discovered Alien Bases on the Moon II*, author Fred Steckling writes about astronauts witnessing extraterrestrial craft on the Moon during the *Apollo 11* mission. He claims that images containing moon anomalies and UFOs from the missions were airbrushed out before being released to the public. Several of the astronauts have reported witnessing something anomalous in space. Many feel that due to the astronauts training, experience, and credibility, that they believe the stories of the strange occurrences.

Apollo Missions 1 – 6

Apollo 1 was the first manned mission of the Apollo program. It was to be the first crewed mission and was looked forward to with great anticipation. It was scheduled to launch on February 21, 1967. The crew included Command Pilot Virgil "Gus" Grissom; Senior Pilot Ed White; and Pilot Roger B. Chaffee. The mission never took place. On January 27, 1967, during a pre-flight test, a flash fire quickly spread through the *Apollo 1* Command Module. Sadly, all three astronauts perished. It was a sorrowful and tragic time at NASA in the days following the disaster. NASA was intent in recovering and continuing the Apollo program. There were no space missions specifically named Apollo 2 through 6. These

The *Apollo 1* crew, Vigil Grissom, Ed White, and Roger B. Chaffee. (NASA)

were missions that were unmanned and designed to test different aspects of the program.

Apollo Mission 7

Apollo 7 was the first manned mission after *Apollo 1*. *Apollo 7* launched on October 11, 1968. The crew consisted of Commander Walter Schirra Jr.; Lunar Module Pilot R. Walter Cunningham; and Command Module Pilot Donn F. Eisele. It was the first Apollo mission to transport astronauts into space. It was a step closer to reaching the goal of one day placing men on the Moon. It was on this mission that the first strange incident took place in space during an Apollo mission. It had been a successful launch. The crew orbited the Earth 163 times and spent 10 days and 20 hours in space. At some point during the mission, the crew noticed something truly spectacular, if not downright eerie. They saw a large UFO that has been variously described as "angelic," "beautiful," "large" and "metallic." The *Apollo 7* crew took several photographs of the object that showed what could only be some sort of sophisticated

211

The *Apollo 7* crew, Walter Cunningham, Walter Schirra Jr., and Donn F. Eisele.

spacecraft. What came next is quite telling as there appears to have been an attempted cover-up of what the astronauts had seen and photographed.

According to the story, someone sought to cover up the UFO shown in the photographs with duct tape. Whereas the astronauts certainly needed photos from the first manned Apollo mission to fly into space, they had not bargained on encountering something out there, and someone unnamed appears to have made a rudimentary attempt to conceal it. Strangely, on one of the images the UFO can still be seen. It appears that this image was accidently missed in the duct tape cover-up and somehow slipped through unnoticed. One can only wonder if this was done purposely. Was someone attempting to get a message out that the astronauts had witnessed something extraordinary while out there? Whatever it was that they photographed can clearly be seen in the picture today. It has been speculated that whatever the mysterious object is, it was made by highly advanced beings that are far superior to humans. UFO proponents believe that the Earth's space programs are being watched, and that extraterrestrials are especially interested in human achievements when it comes to space travel. This is speculated to have been the reason behind this UFO sighting. They may have been watching the astronauts, or knew they were coming and wanted to be seen. If this is true, it begs the question, *"Are there Moon operatives on Earth that knew when the Apollo 7 astronauts would be launching?"*

Image of a UFO taken by the crew of *Apollo 7*.

Apollo Mission 8

Apollo 8 launched on December 21, 1968 with the goal of reaching lunar orbit. On this mission the astronauts worked in the technical and scientific areas. The crew consisted of Commander Frank Borman; Lunar Module Pilot William Anders; and Command Module Pilot James Lovell. It was during this mission that Anders took the now famous "Earthrise" photograph on December 24,

The *Apollo 8* crew, Frank Borman, James Lovell, and William "Bill" Anders.

213

1968. During the *Apollo 8* mission, the astronauts experienced several weird, perplexing events. During their journey, the crew was astonished to see a shining spherical-shaped UFO moving rapidly through the darkness of space. The astronauts reported that the object emitted a light that was so brilliant they could scarcely see the inside of their own ship. As the UFO passed them, their ship pitched and yawed to the point that the men nearly lost control of the craft. At that point, Lovell contacted ground control stating, "We have been informed that Santa Claus does exist." Some suggest that the term "Santa Claus" was code for a UFO or extraterrestrial vessel. The fact that the astronauts and ground control had developed code words lets us know that the topic of UFOs was discussed before the astronauts went up.

Apollo Mission 9

Apollo 9 commenced on March 3, 1969. It was the Apollo program's third manned mission. The crew was to implement crucial procedures for landing men on the surface of the Moon and testing the Lunar Module in space for the first time. The Lunar Module would eventually transport astronauts in later missions to the surface of the Moon. It would return them to the Command Module once the work on the surface of the Moon was completed. The crew consisted of Commander James McDivitt; Command

The *Apollo 9* crew, James McDivitt, David Scott, and Rusty Schweickart.

Module Pilot David Scott; and Lunar Module Pilot Rusty Schweickart. It was on March 10, 1969 that the crew witnessed something strange in space. It was to be one of the first reported sightings of the large, strangely-shaped objects that we commonly refer to these days as a cylindrical-shaped ship. The crew witnessed one of these objects crossing the moon. Reportedly, the astronauts photographed it. There have been several reports of cylindrical-shaped UFOs seen in the skies over Earth. Ufologists speculate that what the astronauts saw that day traversing the Moon was extraterrestrial related. These UFOs are seen so often that I wonder, *"Are these cylindrical-shaped objects seen on the Moon connected to the many that have been witnessed near Earth? And What is their purpose?"*

Apollo Mission 10

Apollo 10 was the final mission before the history-making *Apollo 11* that would place men physically on the Moon. It was the Apollo program's fourth manned mission, and the second time that astronauts would orbit the Moon. One could call it a kind of test run or dress rehearsal for the *Apollo 11* astronauts. The crew included Commander Thomas Stafford; Lunar Module Pilot Eugene Cernan; and Command Module Pilot John Young. *Apollo 10* experienced a problem when the Lunar Module first separated from the Command Module. It would be the Lunar Module that would transport Neil Armstrong and Buzz Aldrin to the surface during the *Apollo 11* mission, so, it was critical that things run smoothly during the *Apollo 10* operation. In this operation, the Lunar Module was to stop just short of landing on the Moon. As they neared the surface, the crew inside the capsule needed to get a lock on the signal from the Command Module. This would allow the astronauts to maintain a precise altitude above the surface. However, they were unable to achieve this goal due to a transponder failure on the Command Module. Without a lock, the entire mission was at risk. Subsequently, the crew in the Lunar Module became stranded. The astronauts attempted manual override and failed. They contacted ground control for advice with no success. During the commotion, Cernan noticed a UFO near the craft. The object quickly disappeared. Within the next

215

The *Apollo 10* crew, Eugene Cernan, John Young, and Thomas Stafford.

few moments, the conditions within the Lunar Module improved. The radar systems began to work, and the astronauts were able to secure a lock. Ground control then gave them the go ahead to continue toward the surface.

Once the astronauts returned home, a video that Cernan had taken of the UFO was evaluated. There were various opinions among officials as to what the astronauts had seen. Many believed that what had happened with the UFO passing and then the systems operating afterward was merely a coincidence. Others speculated that the UFO appearing at a time when the astronauts were having difficulties was no accident, and that it was benevolent extraterrestrials that aided the crew by mysteriously repairing the stalled equipment. Interestingly, there were some that maintained that the UFO was attempting to warn the astronauts away by disrupting the operation. One plausible explanation that was brought forward is the possibility that the UFO was emitting signals that were interfering with the equipment on the Lunar Module. What happened with the *Apollo 10* mission and the UFO is still considered a mystery today. The question is, *"Did extraterrestrials, in a way unknown to us, assist the astronauts when they became stuck in space?"*

Apollo Mission 11

The *Apollo 11* mission was one of the most anticipated events in world history. It is without a doubt one of the greatest accomplishments of man. *Apollo 11* was launched on July 16, 1969. It was the *supreme* mission, as it would be the first time that humans walked on another world. The crew consisted of Commander Neil Armstrong; Command Module Pilot Michael Collins; and Lunar Module Pilot Buzz Aldrin. It was also this mission that would determine the continuation of future Apollo missions. If successful, there would be more missions of its kind, with the goal of one day establishing a lunar colony. The launch into space was successful. NASA and the public were relieved that the crew had made it safely to space. The operation appeared to be running smoothly to the audience back on Earth. However, unbeknownst to the public as well as mission control, the astronauts were witnessing something unusual in space. The men saw what they could only identify as a UFO, and it appeared to be following them. In his autobiography, *Return to Earth,* Buzz Aldrin described it as having the appearance of a brightly-lit letter "L." After some discussion as to how to proceed, the astronauts eventually contacted ground control, but out of caution did not tell the ground personnel that they were being trailed by an UFO. Instead, they decided to ask about the location of the Saturn V and its proximity to them. They were told that the Saturn V was approximately 6,000 miles away from their position. It has been speculated that what the astronauts saw that day was a part of the *Saturn V*, or some other piece of equipment. Others maintain that it was quite possibly an extraterrestrial spacecraft. Thus far, there is no concrete answer as to what was seen outside the window that day by the astronauts. *Were extraterrestrials following Apollo 11 to the Moon that day?*

Once they landed on the Moon, as the now famous story goes, Neil Armstrong and then Buzz Aldrin stepped onto the lunar surface. Armstrong proclaimed, "That's one small step for man, one giant leap for mankind." What is said to have happened next was not seen on video. It was not heard on the broadcast, but according to what has now become almost a legendary tale, the astronauts witnessed something that can only be described as shocking, disturbing, and frightening. As the story goes, the astronauts saw extraterrestrial spaceships sitting on the edge of a crater. Purportedly, Armstrong conveyed a message to mission control via a private medical line stating that there were extraterrestrial spaceships parked on the rim of a crater, and that

The *Apollo 11* crew, Neil Armstrong, Michael Collins and Edwin "Buzz" Aldrin.

they were being watched. When the astronauts were reporting the situation in which they found themselves to mission control, there was a two-minute radio silence so that the public would not know there was a problem. The conversation was overheard by ham radio operators who had their own VHF receiving facilities that bypassed NASA's broadcasting outlets. The following is the exchange purportedly picked up between Armstrong and mission control.

> **Armstrong**: What was it? What the hell was it? That's all I want to know!
> **Mission Control**: What's there? Malfunction (garble). Mission Control calling Apollo 11
> **Apollo 11**: These babies are huge, sir!.... Enormous!... Oh God! You wouldn't believe it!... I'm telling you there are other space-craft out there... lined up on the far side of the crater edge!... They're on the Moon watching us!

Interestingly, in his book *Visitors from Other Worlds* (pg. 56), Author and Researcher Brad Steiger writes about Dr. Sergei Bozhich, who had witnessed the Russians observing the *Apollo 11*

Moon landing. Steiger stated that in Bozhich's opinion, the UFOs seemed poised to help the two astronauts should something have gone wrong with the Moon landing. Bozhich insinuated that after the *Eagle* (Lunar Module) had set down and everything appeared secure, that the extraterrestrial ships left. One wonders, if the astronauts had run into difficulties when landing on the Moon, would extraterrestrials have assisted them?

Apollo 11's Smoking Gun

Christopher Kraft (1924-2019) was an aerospace engineer. He worked as a director of the NASA tracking base in Houston during the Apollo program and worked very closely with *Apollo 11*. He has been referred to as the "legendary" founder of NASA's mission control. After leaving NASA, Kraft recounted what happened on the Moon with the *Apollo 11* astronauts. The following is from Researcher and Writer Steve Omar's article titled "U.F.O. and Reported Extraterrestrial on Moon and Mars."

ASTRONAUTS NEIL ARMSTRONG and BUZZ ALDRIN speaking from the Moon: "Those are giant things. No, no, no… this is not an optical illusion. No one is going to believe this!"
MISSION CONTROL (HOUSTON CENTER): "What… what…what? What the hell is happening? What's wrong with you?"
ASTRONAUTS: "They're here under the surface."
MISSION CONTROL: "What's there? Emission interrupted… interference control calling *Apollo 11*."
ASTRONAUTS: "We saw some visitors. They were there for awhile, observing the instruments."
MISSION CONTROL: "Repeat your last information."
ASTRONAUTS: "I say that there were other spaceships. They're lined up on the other side of the crater."
MISSION CONTROL: "Repeat… repeat!"
ASTRONAUTS: "Let us sound this orbita… In 625 t… automatic relay connected… My hands are shaking so badly I can't do anything. Film it? God, if these damned cameras have picked up anything… what then?"

219

MISSION CONTROL: "Have you picked up anything?"

ASTRONAUTS: "I didn't have any film at hand. Three shots of the saucers or whatever they were that were ruining the film."

MISSION CONTROL: "Control, control here. Are you on your way? Is the uproar with the U.F.O.s over?

ASTRONAUTS: "They've landed there. There they are and they are watching us."

MISSION CONTROL: "The mirrors, the mirrors… have you set them up?"

ASTRONAUTS: "Yes, they're in the right place. But whoever made those spaceships surely can come tomorrow and remove them. Over and out."

I refer to this conversation as the "smoking gun," because it is doubtful that someone with Kraft's credentials and reputation, who played such an important role in NASA, would come forward with a falsehood regarding the mission.

Armstrong and Aldrin worked on the lunar surface for three hours collecting samples, setting up the flag and leaving a plaque and other items. It was a historic day, and for any lunar inhabitants, that day may have been a first for them as well. It was the day that Earthlings first set foot on their world. The question is, "were they happy about it?"

The next five missions to the Moon were far from routine. Although it is said that the public lost interest in the missions, one can only wonder how people would have reacted had they known some of the strange events that occurred during the Apollo missions. If the story of the *Apollo 11* astronauts encountering extraterrestrials while sitting in their spacecraft on the Moon is true, someone made an incredible decision to continue with the other missions knowing that there were aliens out there. However, to cancel the missions after *Apollo 11* would have been a signal to the world that something was amiss with the Moon. The sudden cancellation may have frightened the public. By that time, they surely had already spent billions on upcoming missions. Officials more than likely felt as though they had no choice but to continue. Also, it may have appeared that the extraterrestrials were friendly.

"Perhaps."

When the *Apollo 11* astronauts returned, a press conference was held. There has been a lot of discussion of the astronauts' demeanor during that conference. Were they jovial, ecstatic, happy and full of stories? No. At the conference their manner can be described as distant and despondent. One wonders if the astronauts' behavior that day was due to what they had seen and the burden of keeping it a secret.

In his book titled *Suppressed Inventions & Other Discoveries*, Jonathan Eisen writes about an event that Neil Armstrong attended where he was speaking with an unnamed professor. According to sources, this exchange took place during a NASA symposium. According to Eisen, Armstrong was asked by the professor about his trip to the Moon. He was asked directly about the events that he had experienced there. Armstrong relayed to the professor that they (he and Aldrin) had seen spaceships that were highly advanced and that they were huge. He also stated, according to this story, that they were "warned off the Moon." According to the conversation, Armstrong indicated that he and the others knew that there was always a chance of encountering something or someone out there. The following is allegedly an account this exchange. It has been circulated in various websites and publications for years.

Professor: What REALLY happened out there with *Apollo 11*?
Armstrong: It was incredible, of course we had always known there was a possibility, the fact is, we were warned off! [by the Aliens]. There was never any question then of a space station or a moon city.
Professor: How do you mean "warned off"?
Armstrong: I can't go into details, except to say that their ships were far superior to ours both in size and technology – Boy, were they big!… and menacing! No, there is no question of a space station.
Professor: But NASA had other missions after *Apollo 11*.
Armstrong: Naturally NASA was committed at that time, and couldn't risk panic on Earth. But it really was a quick scoop and back again.

Armstrong confirmed that the story was true but refused to go into further detail, beyond admitting that the CIA was behind the cover-up. Today some of the audio tapes and images from the *Apollo 11* missions (as well as others) are said to be missing or in some cases "accidentally" destroyed. We may never know the truth of all that happened during that mission to the Moon. However, with Christopher Kraft's revelation, and the fact that at one time he was an integral part of the Apollo program and very well-respected, then perhaps we can be confident in the so-called rumors about the *Apollo 11* astronauts seeing ships, and the fact that there are (or were) other worldly-beings on the Moon. There is also a tale that has been circulating for years stating that when one of the astronauts opened the door of the Lunar Module, he immediately saw an extraterrestrial. The being is said to have had an ethereal appearance. *Did an extraterrestrial come to meet with the Apollo 11 astronauts?* If there was such an encounter, that could have been the point when they were "warned off the Moon."

Apollo Mission 12

Apollo 12 was the first mission after the history-making *Apollo 11*. It was launched on November 14, 1969. Its destination was the Ocean of Storms (Oceanus Procellarum), the largest dark spot on the Moon. The team included Commander Charles Conrad Jr.; Command Module Pilot Dick Gordon; and Lunar Module Pilot Alan Bean. From the start, *Apollo 12* was inundated with inexplicable, perplexing occurrences. In his book *Our Mysterious Spaceship Moon,* Ufologist Don Wilson writes, "There is widespread agreement by researchers (unquestionably backed up by authenticated evidence from NASA files) that mysterious, unexplainable things did happen on the expedition of *Apollo 12*." The first incident occurred 30-seconds after takeoff, when a lightning bolt hit the Saturn V launch rocket. Thirty-seconds later, a second bolt struck it again. All power was lost for three minutes. For a period, there was a question as to whether *Apollo 12* would continue or if the mission would be aborted. Even more mysteriously, reports came in from observatories around Europe asserting that there were two UFOs with flashing lights pacing the Saturn V. The next day, the crew confirmed to NASA that they too had observed UFOs following their craft. One of the objects was reportedly rotating as it traveled. In due course, the UFOs took off speedily into the darkness and disappeared. The astronauts later stated that they believed that whoever was controlling the UFOs were benevolent. Ufologists maintain that as with the other Apollo missions, extraterrestrials were watching the launch and were there to observe the mission. Interestingly, others have speculated that the lightning bolts were purposely sent as a warning for the astronauts to halt their activity.

The *Apollo 12* crew, Charles Conrad Jr., Richard "Dick" Gordon, and Alan Bean.

223

Once the *Apollo 12* Lunar Module landed safely on the surface, the astronauts went about fulfilling their assignments. However, strange occurrences continued to follow the team. An image that was taken during the *Apollo 12* mission, and one that has received attention from ufologists and lunar researchers, is NASA photograph No. AS12-497319. In the picture, there appears to be a large UFO hovering above one of the astronauts as he is working on the Moon. It appears that he was being monitored. There is a strange tale surrounding the *Apollo 12* mission that is very odd and a bit eerie at the same time. However, the story is said to be completely true. As the story goes, as the astronauts went about their work on the lunar surface, they were astonished to see a semi-translucent object in their proximity. They described it as a shimmering pyramid that emanated the colors of the rainbow. Weirdly, this object was suspended just above the surface. The

Alan Bean of *Apollo 12*. In Bean's helmet the reflection of Charles Conrad can be seen. In the background, there is a UFO hovering in the distance.

astronauts felt that they were being carefully observed by the object! *Could it have been a camera or probe? If there are extraterrestrials on the Moon, how much data do they have about humans? Do they have images and recordings of us? Do they study our cultures and languages?*

While in space, the crew as well as ground control heard incomprehensible, perplexing sounds coming from the radio. The source could not be located. Could this have been coming from the same source as heard by the *Apollo 8* astronauts? Was there someone out there trying to communicate with the astronauts again? On the return trip, while in proximity of Earth, *Apollo 12* ended its journey as it begun during the launch, with a UFO sighting. Reportedly, again, there was a UFO spotted near the craft. It too had a flashing light. As with the others, it eventually vanished. In all cases of the mysterious sightings and unexplained sounds that were linked to *Apollo 12*, no rational explanations were ever given.

Apollo Mission 13

Apollo 13 launched on April 11, 1970. It was scheduled to be the third mission to land on the Moon. The crew comprised Commander James A. Lovell Jr.; Command Module Pilot John L. Swigert Jr.; and Lunar Module Pilot Fred W. Haise Jr. Many people are aware that the *Apollo 13* mission ran into difficulties. Due to a break in the oxygen tank on the Service Module, there was an explosion on board the craft, leaving *Apollo 13* crippled in space. Unable to land, the astronauts orbited the Moon instead. Consequently, *Apollo 13* returned to Earth without finishing the mission. As in the case of *Apollo 10,* there is yet another story that suggests that extraterrestrials intervened in an Apollo mission and helped the astronauts.

According to sources, the crew allegedly received a communication from an extraterrestrial spacecraft advising them on how to safely return to Earth. According to this story, without the assistance of these beings the astronauts would not have made it home. Some claim that when they returned to Earth, the crew was told not to reveal the truth about what really happened in space. This story of course cannot be substantiated. There are no NASA

The *Apollo 13* crew, Fred W. Haise Jr., James "Jim" Lovell Jr., and John L. Swigert Jr.

experts or witnesses that have come forward with this. Could this story be true? After everything that we have read to this point, and after all the stories with merit, it is difficult to know for certain. In any event, it makes a very interesting tale.

Apollo Mission 14

Apollo 14 was the replacement mission for *Apollo 13*. It was launched on January 31, 1971. Its destination was the Fra Mauro crater. Scientists sought information from this region believing that it held data that would help them to better understand the Moon's past. It included Commander Alan Shepard; Command Module Pilot Stuart A. Roosa; and Lunar Module Pilot Edgar Mitchell. While on the surface the astronauts took pictures that showed something bewildering. Some of their images included strange blue lights. The blue lights were mysterious and fascinating. They have been dubbed "blue lights" and also "blue flames." These very odd lights seem to have been strategically placed to be noticed by the Apollo astronauts during the mission. The lights took on different shapes, and can be found in different points in several photographs. These lights are lit up like bright blue lanterns. They also appear to have an object inside of them. They are seen in the sky near the Apollo astronauts and on the surface. They are so clear and distinct that one wonders if they were a type of clue given to the astronauts that they were not alone there. These same

The *Apollo 14* crew, Edgar Mitchell, Alan Shepard, and Stuart Roosa.

types of lights also appear in *Apollo 16* images. *Were the Apollo 14 astronauts under surveillance?*

Apollo Mission 15

Apollo 15 was the fourth Apollo program mission to land on the Moon. It was the first time that astronauts would drive the Lunar Roving Vehicle. *Apollo 15* launched on July 26, 1971. This operation was more scientifically oriented than the ones prior. On board was Commander David R. Scott; Command Module Pilot Alfred Worden; and Lunar Module Pilot James Irwin. The crew landed in the Hadley-Apennine region, which is known as a dangerously rugged area. The region is inundated with crater-pocked terrain, rocks, and boulders that are difficult to maneuver around. Several inexplicable events are connected to *Apollo 15*. In one account, as they went about their work, the two astronauts were almost hit by an unknown object that whizzed over them. The astronauts had no idea what it was, nor where it came from.

Unfortunately, there is no description of the item, however, it seems as if something was attempting to make its presence known to the crew. Because we only have bits and pieces of these stories, it is impossible to make an accounting of what exactly happened, and what the astronauts in most cases thought. However, just having the knowledge that these strange incident occurred is enough to get

The *Apollo 15* crew, James Irwin, David R. Scott, and Alfred Worden.

the conversation going and make a case for further investigation. Additionally, there are anomalous objects in images taken by the *Apollo 15* astronauts. One photograph shows a cylinder-shaped object located on top of the rim of a crater, in the southern area of the Moon's far side. It was situated near the Delporte-Izsak region. The object is believed to be an ancient spaceship. It was featured in the *Apollo 20* hoax, where it is said a secret mission had been sent to the Moon to investigate what was thought to be an ancient ship. If it is a ship, then it is an enormous one. In the pictures, it is situated on top of the lunar surface rather than blending into it and therefore is easily discernable.

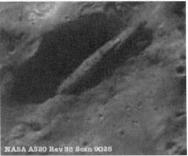

A mysterious cylinder-shaped object from an image taken by the *Apollo 15* astronauts.

Apollo Mission 16

Unlike *Apollo 15,* the *Apollo 16* astronauts experienced mysterious phenomena before landing on the Moon. *Apollo 16* was launched on April 16, 1972. Its crew included Commander John Young;

Command Module Pilot Thomas Kenneth Mattingly; and Lunar Module Pilot Charles M. Duke, Jr. Their goal was the Descartes Highlands, an area filled with hazardous terrain that included mountains, craters and what is interestingly described as gleaming white and pink boulders, which are not something one would expect to find on the lunar surface. Reportedly, the astronauts saw bright flashes of light while in orbit but were unable to locate the source. Later, as Ken Mattingly went about his duties aboard the Command Module while the others were on the surface, a brilliant light appeared out of nowhere, and streaked across the sky. It disappeared on the far side of the Moon. As had become the norm, the astronauts' stories confounded NASA scientists. Several explanations were given for these two incidents. One theory was that of a micrometeorite striking the surface of the Moon. Some thought that the lights were caused by cosmic rays that may have been affecting the men's eyes. However, these hypotheses were

The *Apollo 16* crew, John Young, Thomas Kenneth "Ken" Mattingly, and Charles "Charlie" Duke, Jr.

eventually rejected. It was concluded that there were no logical explanations for what the crew had experienced in orbit and on the surface.

Apollo Mission 17

The Apollo program's final mission to the Moon was *Apollo 17*. It was launched on the night of December 7, 1972. The goal was the Taurus-Littrow valley. The crew was made up of Commander Eugene Cernan; Command Module Pilot Ronald Evans; and Lunar Module Pilot Harrison Schmitt. Their mission was to survey the Moon, photograph areas of significance, and run tests on equipment. It didn't take long before the *Apollo 17* crew began to have extraordinary experiences in space. Although we will never know the extent of everything the astronauts saw out there, the information that has leaked out is stunning and revealing. As I have said, it did not take us long before we ran into something or someone in outer space. During orbit, as with other missions, the crew began to experience very bright flashes of light coming from the vicinity of the Moon, and shining into

The *Apollo 17* crew, Harrison Schmitt, Ronald Evans, and Eugene Cernan.

their craft. They attempted to locate the source of the light but, as with the other missions, were unsuccessful. Later, the crew took photographs that showed anomalous lights and objects.

There are many stories associated with the Apollo missions that can be related to sci-fi; however, the story of the possible Stargate on the Moon is one of the most relatable. In NASA image AS17/AS17-151-23127 there is an odd object that for some sci-fi fans will seem familiar. No one knows for sure what it is that was photographed there, however the object has sparked some very interesting conversations. The image shows a peculiar, bright object located near the lunar surface. It has been likened to a "space portal."

In an article titled "Did Apollo 17 find a Stargate on the

An image (AS17-151-23127) taken by the *Apollo 17* astronauts. It is believed by some to be a possible stargate.

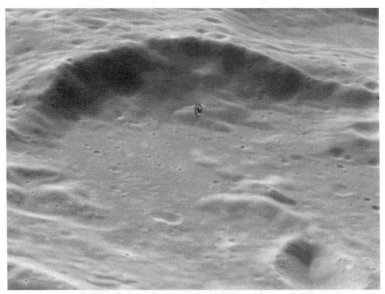

A close-up of the stargate on image AS17-151-23127 taken by the *Apollo 17* astronauts. The dark circular object is blue in the photo.

Moon," Researcher and Author Michael Salla writes, "The object appears to be a space portal of some kind with an eerie blue glowing ring around a central darker portion." Amazingly, it really does resemble the travel device in the sci-fi series *Stargate SG-1*. Although it may appear to be a far out claim, we have already read here about portals and gateways. *Could this be a gateway used by extraterrestrials to come and go on the Moon?* It really is a fascinating subject that I promise to cover in a future book.

A truly startling and eerie incident happened while the *Apollo 17* astronauts were exploring the Moon's surface in the Lunar Rover. As the two were traveling through the Taurus-Littrow valley in the Lunar Rover, NASA was controlling the camera from the ground as the mission was being televised. Suddenly, there appeared an incredible sight in front of the camera that seemed to be an enormous, rectangular-shaped structure. Broadcaster Walter Cronkite, who was covering the mission for *CBS News,* saw it and stated excitedly, "That looks like a manmade structure!"

What the astronauts and the viewers had seen looked like an immense, windowless, rectangular or shoe box shaped structure. The newsfeed was cut, and footage that was shot earlier in the day was replayed on the screen. Millions of viewers witnessed this astonishing event. Twenty minutes later, once the newsfeed was back up, there was an explanation for what had been seen by those watching. Cronkite stated that the Lunar Rover (which carried a camera) had taken a picture of itself! The structure appeared to be the same color as the lunar floor, and even looked to have been made from the same material, sort of giving it a camouflaged look, and blending it into the lunar landscape. Today, there are those who still remember seeing a structure on the Moon when they were young, as they watched the last Apollo mission. Perhaps this incredible viewing was one reason NASA stopped sending men to the Moon. We can only speculate as to what else they saw while there. *What else might they have seen during their ride on the Lunar Rover? Did the astronauts encounter aliens that warned them off the Moon that day?* Some believe so. (A link to the video of this event can be found in the video portion of the bibliography of this book.)

Eugene Cernan gave a small speech during the final hours of the trip. Some have wondered if Cernan was referring to something in particular when giving the speech, which had an air of mystery to it. In saying goodbye to their time on the Moon and alluding to the fact that it was the last Apollo mission, Cernan stated, "I'm on the surface; and, as I take man's last step from the surface, back home for some time to come—but we believe not too long into the future—I'd like to just [say] what I believe history will record. That America's challenge of today has forged man's destiny of tomorrow. And, as we leave the Moon at Taurus-Littrow, we leave as we came and, God willing, as we shall return, with peace and hope for all mankind. Godspeed the crew of *Apollo 17*." Some wonder if there was a hidden message in Cernan's speech, and if he was trying to convey a hidden meaning about the strange things they had encountered in space.

Apollo Missions 18-20

Apollo 18–20 were Apollo program missions that were in the planning stages but were never completed. In 1972 the program was abandoned. *Is it possible that the astronauts encountered extraterrestrials on the Moon that did not want us there?*

Apollo Mission 20, the Hoax

In 2007 an extraordinary story materialized about a mission to the Moon that was said to be *Apollo 20*. As far as we know, the final mission to the Moon was *Apollo 17*. NASA had planned three other missions, which would have included *Apollo 18, 19* and *20*, which were said to have been well into the development stages, with even some production having been done. As we have already established, they were eventually cancelled in 1972. According to the story, *Apollo 20* launched in August 1976 from the Vandenberg Air Force Base. Allegedly, NASA officials became interested in a photograph that was taken by the *Apollo 15* crew. The image showed what appeared to be a mysterious cigar-shaped object situated near the Delporte crater on the far side of the Moon. Officials purportedly believed it to be an ancient spacecraft that measured approximately two and a half miles long (4 kilometers).

According to the story, it was decided to send a clandestine mission to the Moon to investigate it. This mission was said to have been a collaboration between the United States and the Russian space programs. Once the astronauts reached the far side, they found the remnants of an extraterrestrial city, and, as anticipated, a massive, cigar-shaped, miles-long, ancient alien spaceship that has been referred to as a huge mothership. The astronauts boarded the craft and famously discovered the body of an extraterrestrial humanoid female that apparently was the pilot of the craft. She appeared to be in stasis appearing neither alive nor dead, and had apparatuses attached to her eyes and nose that looked as if they were used for navigation. According to the story, the crew detached the extraterrestrial from the apparatus, placed her aboard the Lunar Module, and returned to Earth with her. They named her Mona Lisa.

The perpetrator of this rather elaborate hoax was a mysterious figure by the name of William Rutledge. He posted videos online of the hoax in 2007. He claimed to have been one of the astronauts on a clandestine *Apollo 20* mission. According to Rutledge, *Apollo 20* really happened. He stated that it was a secret mission that was concealed from the public. He said that what they found on the Moon was indeed an ancient alien spacecraft. He named the other astronauts as Leona Marietta Snyder and Alexei Leonov. Rutledge

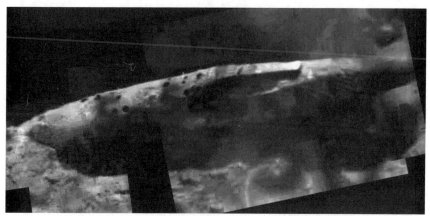

A close-up of the spaceship from the *Apollo 20* "hoax" footage.

stated that he is a United States citizen who is retired and residing in Rwanda, Africa. According to sources, William Rutledge and Leona Marietta do not appear on the NASA personnel lists. Alexei Leonov was a Soviet cosmonaut famous for being the first person to perform a spacewalk on March 18, 1965. The spacewalk lasted 12 minutes and 9 seconds. He was obviously very capable and well qualified to travel to the Moon.

In images taken by the crew of *Apollo 15* in 1971, one can see what appears to be a cylindrical-shaped spaceship in several of the photos, sitting on the rim of a crater, separate from the geological landscape. The image of the ship in Rutledge's photos and that of *Apollo 15* are nearly identical. One could argue that they are one and the same. Interestingly, the ancient spaceship in the *Apollo 20* videos show unknown foreign lettering and symbols. It seems that someone went through a lot of trouble to create them, to make the ship look authentically alien. What is troubling about this story is that the details seen in the videos that Rutledge claimed he took on the Moon look genuinely authentic.

From the details inside the craft right down to the mission insignia, everything looks as if it were from the Apollo program era. It appears that someone went through a lot of trouble and expense to pull this hoax off. It has even been commented that whoever put the videos together knew intimate details of the designs of a lunar module, equipment, and technology, along with the intricacies and complexities of it. It certainly appears that if *Apollo 20* was a hoax, then it was a brilliant and expensive one. If it was not, then it was a secret mission that has been covered up.

Eventually, as authentic as the story appears, it was debunked. There was a two-fold reason. An Apollo mission needed a Saturn V rocket to succeed. No Saturn V was launched from the location given by Rutledge after *Apollo 11*. In addition, a French sculptor by the name of Thierry Speth came forward and claimed responsibility for creating the story and artwork in the video. While some have questioned whether Speth was being truthful in his taking credit for the artwork in the video, it is difficult to argue that a Saturn V was launched. However, given the many theories and speculations about secret missions to the Moon, is it possible that this story could be true?

Warned off the Moon?

The tale of the astronauts being warned off the Moon has become legendary. The story has been circulated for many years. It is said that the reason the Apollo program ended was because the astronauts were "warned off the Moon" by extraterrestrials. These otherworldly beings are believed by some to have been encountered by astronauts and conveyed that the Earthlings were not welcome and were not to return. Supposedly, it was the *Apollo 11* astronauts that received this message. As we know, after *Apollo 11* NASA did send astronauts to the Moon for several more missions. The *Apollo 17* mission was the last of the NASA trips to the Moon. *Apollo 17* has also been rumored to have received such a message. Some believe that the alleged moon inhabitants considered the astronauts to be intruders. Conspiracy theorists believe that the astronauts being warned off the Moon is the reason that the missions were halted. There is also a story that has been circulated that not only were the astronauts warned off the Moon, but that they were escorted off.

Could this be the key to the picture from a photograph taken during the *Apollo 12* mission (NASA No. AS12-497319), where what appears to be a UFO is hovering above one of the astronauts as he is working on the Moon? Were the astronauts being monitored? This idea lends credence to the opinion of some scholars that have proposed that Earth could be under surveillance by advanced extraterrestrials. In the *UFO Investigator* of December 1958,

An image (AS12-497319) taken during the *Apollo 12* mission of a UFO
hovering over one of the astronauts.

Professor Harold D. Lasswell of the Yale Law School stated, "The
implications of the UFOs may be that we are already viewed with
suspicion by more advanced civilizations and that our attempts
to gain a foothold elsewhere may be rebuffed as a threat to other
systems of public order."

In exploring the reasons why an alien race would warn us off
the Moon, several ideas have been put forward.

1. There was an ancient alien war, and the agreement was for
us to stay away.
2. They see us as an immature, warring race with violent
tendencies.
3. They do not want to interact with aliens (us).
4. They consider the Moon their home and we were trespassing.

237

5. There is a galactic directive in place, and they are not permitted to interact with us, so as not to interrupt our natural evolution.

6. We disturbed them by trespassing, bombing the Moon, and leaving trash on the surface of the Moon.

If Professor Lasswell is correct, and if we really were warned off the Moon, what does this mean for future space travel?

The Magnificent 12

Twelve men have walked on the Moon. These first human space explorers have set the stage for the next group of people to explore the stars. In their youth, they could not have imagined that they would one day walk on another world and be hailed as heroes for doing so. Neither would they have imagined that in their lifetime they would encounter beings from outer space, and that the subject of walking on the Moon, extraterrestrials, UFOs and more, would be associated with their names. These men are not just the first men to walk on another cosmic body, they are the first to have extraterrestrial encounters in outer space. We of course do not have all the details of what happened out there. We have only bits and pieces of what they encountered. However, the experiences and views of these men make up part of the fabric of our galactic history with the Moon. It is worth looking back at some of their encounters and listening to their words from all those years ago about their missions to the Moon. They include Neil Armstrong, Edwin "Buzz" Aldrin, Charles "Pete" Conrad, Alan Bean, Alan B. Shepard Jr., Edgar D. Mitchell, David R. Scott, James B. Irwin, John W. Young, Charles M. Duke, Eugene Cernan, and Harrison H. Schmitt.

Edwin (Buzz) Aldrin

Buzz Aldrin (1930—) was the Lunar Module Pilot on the history-making *Apollo 11,* and the second man to step foot on the Moon. Before becoming an astronaut Aldrin served as an Air Force colonel and a pilot. In addition, he holds a doctorate of Science in Astronautics. Aldrin also served onboard *Gemini 12* where he set a record for extravehicular activity. On July 20, 1969, as he stepped

President Barack Obama hosts Neil Armstrong, Michael Collins, and Edwin "Buzz" Aldrin in the Oval office at the White House on July 20, 2009.

onto the Moon, he famously described the ominous scene around him, saying, "Beautiful, beautiful, magnificent desolation."

In his first autobiography, *Return to Earth*, Aldrin wrote of seeing a UFO during the trip to the Moon, stating, "There was something out there that was close enough to be observed." The UFO has been described as luminous and L-shaped, and it was pacing their spacecraft. In a statement from an interview in 1969, Aldrin commented on that same UFO sighting and how the astronauts decided to carry on with the mission, even though there was something strange outside. Aldrin stated, "Now, obviously, the three of us were not going to blurt out, 'Hey Houston we got something moving along side of us and we don't know what it is.'" There is a theory that extraterrestrials were following this anticipated mission to the Moon. From what we know of this story, we can rightly deduce that apparently these extraterrestrials were waiting for them when the men stepped onto the lunar surface.

Neil Armstrong

Neil Armstrong (1930–2012) was the Commander of the legendary *Apollo 11* mission to the Moon that would first place men on the lunar surface. Armstrong had an illustrious career in several areas including being an aeronautical engineer, a naval aviator, a test pilot, the command pilot for the *Gemini 8* mission,

MOON ALIENS

Riddle of two UFOs in crater as Apollo made historic landing

WERE aliens watching when US astronauts first landed on the Moon? Astonishing claims from both American and Russian sources say that two UFOs were already on the surface.

Sunday Mirror Reporter

Apollo spaceship commander Neil Armstrong spotted them as he made his historic "one small step for men, one giant leap for mankind." But then, it is claimed, his radio report of the sighting was blacked out by mission controllers on earth.

NASA—the National Aeronautics and Space Administration — discuss the claims as "absolutely ridiculous." But

Maurice Chatelain, a former NASA consultant, insists his version of the first Moon landing in 1969 is true.

He says Apollo leader Armstrong reported seeing the craft perched on the rim of a crater.

"His reports were blacked out in broadcasts to the world for security reasons," says Chatelain.

Chatelain's claims are backed by two Soviet

space experts who say they learned about the incident two years ago.

But NASA's chief spokesman John McLeaish, who denied the story, said: "The only breaks in transmission from Apollo 11 occurred when it went round the other side of the Moon.

"The only conversations we have never made public were private

talks between the astronauts and doctors."

Chatelain says that NASA ordered a cover-up of the Moon incident.

He went on to claim that while Armstrong was reporting his findings, his colleague Buzz Aldrin filmed the alien craft from inside Apollo 11.

In Moscow, university professor Dr. Vladimir Asbasha said: "I am

certain this episode took place but it was censored by NASA.

Another Soviet space expert, Dr. Sergei Boshich said: "It is my opinion that beings from another civilisation picked up radio signals from earth and then spied on the Apollo landing to learn the extent of our know-how.

"Then they took off without making contact."

News article from the UK *Sunday Mirror Reporter* about a possible sighting of UFOs on the Moon by the *Apollo 11* astronauts.

as well as a university professor. However, he is best known for being the Commander of the *Apollo 11* mission, and for being the first man to walk on the Moon. When the Lunar Module set down on the Moon's surface, Armstrong stated, "The *Eagle* has landed," to let the world know that the trip to the lunar surface was a success. Once he finally took his first step on the lunar surface, he famously stated, "That's one small step for man, one giant leap for mankind." The audience, for what can be referred to as "a groundbreaking momentous occasion" was global. The event was watched by millions of people around the world. That is the Neil Armstrong that we know of and whom many remember. What many do not know is his story after the Moon landing. After the *Apollo 11* mission, Armstrong, according to sources, became something of a recluse. Unlike some astronauts that embraced their fame and success, Armstrong seemed to retreat into life, making few public appearances. It is believed by some that Armstrong experienced something out there that was so life changing, that he spent the rest of his life quietly contemplating those events.

As a young man, Armstrong began his career with dreams of having an outstanding aviation career. Little did he know in his early years, when he was just starting out, how far he would go in the field which eventually took him to the Moon. On a personal level, he was hailed as being a person of many positive attributes. He has been described as extremely principled, self-assured, sharp, and motivated in his pursuits. He also had an insatiable thirst for knowledge and after returning from the Moon, continued his personal journey to learn more. One conspiracy theory suggests that Armstrong's silence proves that there was never any Moon landing. It is said that because he was a man of high morals and integrity, he simply did not want to perpetuate a lie in saying that they had landed on the Moon. Another hypothesis suggests that Armstrong was quiet about the mission because he saw extraterrestrials on the Moon and was instructed not to divulge what he had seen! In 1994, the commemoration for the twenty-fifth anniversary of the *Apollo 11* Moon mission was held at the White House. There Armstrong gave a memorable speech stating: "Today we have with us a group of students, among America's best. To you we say, we have only completed a beginning. We leave you much that is undone. There are great ideas undiscovered, breakthroughs available to those that can remove one of truth's protective layers. There are places to go beyond belief. Those challenges are yours in many fields of not the least is space, because there lies human destiny."

It has been suggested that Armstrong was alluding to something particular with these comments. They appear to almost be a riddle. *"There lies human destiny."* Could Armstrong's comment be connected to the anomalous events on the Moon that he and the others were not allowed to disclose to the public? You may recall that *Apollo 11* is rumored to have seen ships on the edge of a crater and reportedly received a message for humans to stay away from the Moon. Is it possible, that Armstrong's comments and demeanor were connected to such a secret? A journalist interviewed a close friend of Armstrong after the Apollo mission, confided that Armstrong had told him that when they were descending towards the lunar surface, they saw three enormous silver ships sitting on the crater's perimeter. According to this story, Armstrong deliberately turned at the last minute, so as not to show the audience what the astronomers were seeing. Armstrong passed away on August 25, 2012 from heart failure. Many believe that he took secrets about the Moon with him when he died. Today, Neil Armstrong is honored as an historical figure, one that will always be remembered for the heroics of being the first person to walk on the Moon.

Alan Bean

Alan Bean (1932–2018) served aboard *Apollo 12* as the Lunar Module Pilot. He also served as the Commander of *Skylab 3*. He was the fourth person to walk on the surface of the Moon. Bean is the astronaut in the photograph mentioned above with the strange geometrically-shaped edifice in the background. In June of 1981, Bean left NASA and became an artist, taking up painting fulltime. It was his desire to portray in art what he had witnessed in space and on the Moon. His work has given the public an intimate view of the Moon as can be seen only through an astronaut's eyes. His visions of his otherworldly experiences on the lunar surface have touched many people and have encouraged more discussion about the Moon, and the cosmos in general. He also added color to his paintings, showing the world that the Moon is not simply a large dull gray and white rock. When we see his work, we see the Moon through his eyes, and the way he experienced it. Interestingly, there is a conspiracy theory that states that the Moon has more color than we have been told. Could this be the case? Was it Bean's intention to show the world that the Moon has color, after all?

Eugene Cernan

Before working for NASA, Eugene Cernan (1934–2017) worked as an aeronautical engineer and a fighter pilot. His engineer and flight experience undoubtably contributed to his being chosen as the Commander of the *Apollo 17* mission that launched in December 1972. He also carries the distinct title of being the last person to walk on the lunar surface. In total, Cernan participated in three NASA space missions. The others occurred in June 1966, when he served on *Gemini 9A* as co-pilot and in May 1969, when he was *Apollo 10's* Lunar Module pilot. As the *Apollo17* Lunar Module landed on the Moon, Cernan exclaimed, "We is here, man! We is here!" On that assignment he and astronaut Harrison Schmitt studied the Taurus-Littrow area. There are strange tales of the *Apollo 17* astronauts seeing extraterrestrials and extraterrestrial structures while on the Moon's surface. UFO enthusiasts claim that certain quotes from Cernan allude to him witnessing something out of the ordinary in space. They cite his speech from the Moon on his last day when he stated, "I'm on the surface; and, as I take man's last step from the surface, back home for some time to come—but we believe not too long into the future—I'd like to just [say] what I believe history will record. That America's challenge of today has forged man's destiny of tomorrow. And, as we leave the Moon at Taurus-Littrow, we leave as we came and, God willing, as we shall return: with peace and hope for all mankind. Godspeed the crew of *Apollo 17*."

Additionally, Cernan was once quoted as saying, "Maybe the moon can tell us something about the existence of some ancient civilization, not necessarily on earth, nor necessarily on the moon, but possibly within our own universe, and give us insight into what reality is all about." In a *Los Angeles Times* article by Chriss Nicholas titled "Cernan Says Other Earths Exist," Cernan states, "I'm one of those guys who has never seen a UFO. But I've been asked, and I've said publicly I thought they were somebody else; some other civilization…"

Charles Conrad Jr.

Before becoming an astronaut Charles Conrad Jr. (1930–1999) worked as an aeronautical engineer and naval officer. He was a part of several space missions including *Apollo 12, Gemini 5, Gemini 11,* and *Skylab 2*. He was the third man to walk on the Moon, which occurred during the *Apollo 12* mission. As he stepped onto the Moon, Conrad exclaimed, "Whoopie! Man, that may have been a small step for Neil, but that's a long one for me." There is an eerie tale surrounding Conrad and a photograph that he snapped of Alan Bean as they walked on the Moon. In the image, Conrad can be seen in the reflection, in the screen of Bean's helmet. In the background, a mysterious object can be seen in the reflection of the helmet's screen. The object was strangely shaped. It was configured in a geometric arrangement that was inside of an encircling translucent edifice. It is suspended over the lunar surface. One can even see its shadow.

Researchers examining the photograph maintain that the shadow is evidence that the object is real, and that there was no problem with the camera. There is no explanation as to what this was, although looking at the image, it does appear that the astronauts are being watched. However, one would think that the astronauts would have seen the object. By the time the *Apollo 12* mission went up, given all the strange activity with the former missions, perhaps the men knew there was someone there. When they did witness something anomalous, they kept quiet about it. In this case, the object appears to have been accidentally photographed. If the astronauts were under a nondisclosure agreement, there may have been numerous other photographs taken that show that someone is on the Moon. The astronauts quite simply, were not allowed to tell.

Charles Moss Duke Jr.

Charles Moss Duke Jr. (1935–) was a former Brigadier General for the US Air Force. During his time at NASA, he was the Capsule Communicator for *Apollo 11*. Eventually, Duke was selected for the Lunar Module Pilot position for *Apollo 16*. He was the tenth man to walk on the Moon, and at 36 became the youngest human to do so! One of the strangest stories involving the astronauts, and one that is very out of the ordinary, is a dream that Duke had just prior to the Apollo 16 mission. In a way, one could say that this was a case of "déjà vu."

A week before *Apollo 16* was launched, Duke had a mysterious dream where he met his own(twin) self during the mission. It was a dream that he shared with fellow crewmate John Young. The two were to explore the surface of the Moon together. Duke dreamt that the two were on the Moon, traveling upwards in the Lunar Rover, and eventually found themselves at a ridge. In front of the ridge, they saw tracks. Desiring to see where the tracks led, they contacted Ground Control and obtained authorization to follow them. Once they received the go-ahead, they travelled for a period of time before coming across another Lunar Rover. This second Lunar Rover carried two individuals that were identical to Duke and Young. In the dream, Duke was seeing himself and Young in the Lunar Rover.

A few days later, *Apollo 16* was launched. Once on the lunar surface, Duke found himself in the very same situation, and in the very same place as he was in the dream. Here, the two men were traveling north in the Lunar Rover. Duke was steering. Their goal was the

Charles Moss Duke Jr.

North Ray crater. As they navigated over the lunar floor and began to slowly traverse a hill, Duke recognized the area from his dream days before. He saw the same dark, crystallized hill, which was the exact place that he had found himself in his dream. Of course, we cannot understand the relevance of this dream. Who can know such things? If there were a message there, it is unbeknownst to us. However, it certainly is an interesting story to share.

Duke once addressed the conspiracy theory that we never went to the Moon stating, "We've been to the Moon nine times. Why would we fake it nine times, if we faked it?" Duke also made several other comments that gave testimony to how he felt about his experience on the Moon. "The Moon was the most spectacularly beautiful desert you could ever imagine. Unspoiled. Untouched. It had a vibrancy about it and the contrast between it and the black sky was so vivid, it just made this impression of excitement and wonder." About space he commented, "It was a texture. The blackness was so intense."

There was a mysterious conversation between Charles Duke and John Young where one gets the impression that they were witnessing something strange and special…something unknown, as the astronauts seem to have been speaking in code to each other.

Duke: These devices are unbelievable. I'm not taking a gnomon up there.

Young: O.K., but man, that's going to be a steep bridge to climb.

Duke: You got—YOWEE! Man—John, I tell you this is some sight here. Tony, the blocks in Buster are covered—the bottom is covered with blocks, five meters across. Besides the blocks seem to be in a preferred orientation, northeast to southwest. They go all the way up the wall on those two sides and on the other side you can only barely see the out-cropping at about 5 percent. Ninety percent of the bottom is covered with blocks that are 50 centimeters and larger.

Capcom: Good show. Sounds like a secondar …

Duke: Right out here …the blue one that I described from the lunar module window is colored because it is glass coated, but underneath the glass it is crystalline …the same texture as the

245

Genesis Rock ...Dead on my mark.

Young: Mark. It's open.

Duke: I can't believe it!

Young: And I put that beauty in dry!

Capcom: Dover. Dover. We'll start EVA-2 immediately.

Duke: You'd better send a couple more guys up here. They'll have to try (garble).

Capcom: Sounds familiar.

Duke: Boy, I tell you, these EMUs and PLSSs are really super-fantastic!

What were the astronauts talking about here? More anomalous constructions? One wonders. I get the distinct impression that whatever the men were looking at was immediately rendered classified. How many anomalous, strange objects did the NASA astronauts encounter that we will never know about?

James Irwin

James Irwin (1930–1991) was the eighth man to experience walking on the Moon. Before NASA, Irwin served in the United States Air Force as a pilot, and was an aeronautical engineer, as well as a test pilot. Ultimately, he became the Lunar Module Pilot for *Apollo 15*. In his autobiography titled *To Rule The Night, The Discovery Voyage of Astronaut Jim Irwin* (page 60), Irwin expressed his feelings about landing on the Moon stating, "Okay Houston. The *Falcon* [Lunar Module] is on the plain at Hadley."

He continues, "The excitement was overwhelming...We looked out across a beautiful little valley with high mountains on three sides of us and the deep gorge of Hadley Rille a mile to the west. The great Apennines were gold and brown in the early morning sunshine. It was like some beautiful little valley in the mountains of Colorado, high above the timberline... There was the excitement of exploring a place where man had never been before." Irwin was one of the astronauts who was nearly hit by an unidentified object that flew by as he and comrade David Scott went about their duties. Irwin was a Christian who became even more devout after his walking and working on the Moon. He spoke of having an epiphany while there, stating, "Seeing this has

Astronaut James Irwin salutes the flag near the lunar module during *Apollo 15*.

to change a man, has to make a man appreciate the creation of God." About his experience on the Moon, Irwin once commented, "Being on the moon had a profound spiritual impact upon my life." Irwin fervently believed that he had encountered God in space and dedicated his life to telling others about his spiritual encounter on the Moon. In a *New York Times* interview from 1991, he summed up his feelings about his experience on the Moon, stating, "I felt the power of God as I'd never felt it before."

There is a mysterious conversation that was recorded between astronauts David Scott and James Irwin. They seemed to have been witnessing something unexpected and unusual that excited them. From the dialog it sounds like the men are observing an anomalous structure.

> **Scott**: Arrowhead really runs east to west.
> **MC**: Roger, we copy.
> **Irwin**: Tracks here as we go down slope.
> **MC**: Just follow the tracks, huh?
> **Irwin**: Right we're (garble). We know that's a fairly good run. We're bearing 320, hitting range for 413 ... I can't get over those lineations, that layering on Mt. Hadley.
> **Scott**: I can't either. That's really spectacular.
> **Irwin**: They sure look beautiful.
> **Scott**: Talk about organization!
> **Irwin**: That's the most *organized structure I've ever seen*!
> **Scott**: It's (garble) so uniform in width.
> **Irwin**: Nothing we've seen before this has shown such uniform thickness from the top of the tracks to the bottom.

The last line of this dialog indicates that they had seen other anomalous constructions on the Moon as well. To quote Scott, whatever they were seeing sounds *"really spectacular!"*

Edgar Mitchell

Edgar Mitchell (1930–2016) served aboard *Apollo 14* as the Lunar Module Pilot. He was the sixth man to walk on the Moon, exploring the Fra Mauro Highlands area for nine hours. Before becoming an astronaut, Mitchell was a United States Navy officer and aeronautical engineer. In 1970 he was the recipient of the Presidential Medal of Freedom for an attempted mission to the Moon. In his later years, Mitchell was a UFO researcher and author, and founded the Institute of Noetic Sciences. For some, Mitchell was a bit controversial due to his coming forward with his beliefs in extraterrestrials and comments about UFOs. He was moved by his experience on the Moon and completely convinced that we are not alone in the universe as a result. He believed that we are being visited, that someone else is on the Moon and relayed his opinions publicly and often. He was not shy in expressing his thoughts, opinions and ideas and brought a lot of insight and knowledge on

Edgar Mitchell.

the UFO phenomenon and our place in the universe.

Mitchell participated in numerous interviews and several documentaries expressing his ideas, and even though he passed away in 2016, he left a wealth of information for us to review and contemplate. In his book *Earthrise* he tells of his mission to the Moon, and explains how traveling there gave him a different outlook on life, and an understanding that we are not alone in the universe. Mitchell once recalled an experience that he had while working on the lunar surface, stating, "I had to constantly turn my head around, because we felt we were not alone there. We had no choice but to pray." Later, after becoming a UFO proponent, Mitchell once commented, "I have no doubt that extraterrestrials could very well have populated or made structures on the far side of the moon."

Harrison Schmitt

Harrison Schmitt (1935–) was the second to last person to step on the Moon. This was during the final mission to the Moon where Schmitt was the Lunar Module Pilot for *Apollo 17*. Before becoming an astronaut Schmitt worked as a geologist and at one time was a US Senator for New Mexico. He is also the author of *Return to the Moon: Exploration, Enterprise, and Energy in the Human Settlement of Space*. There are several stories involving Schmitt and his comrades from *Apollo 17* that tell of strange experiences and phenomena that they witnessed on their way to the Moon.

There are stories of mysterious lights, a rumor about a lost cosmonaut, a possible extraterrestrial sighting, and a photograph of what some believe may be a stargate. About his experience on the Moon, Schmitt once stated, "It's like trying to describe what you feel when you're standing on the rim of the Grand Canyon or remembering your first love or the birth of your child. You have to be there to really know what it's like."

David Scott

David Scott (1932–) was the seventh person to walk on the lunar floor. Scott was in the third group of astronauts chosen by NASA in October 1963. His missions for NASA included *Gemini 8,* where he served alongside Neil Armstrong; *Apollo 9* where he served as the Command Module Pilot; and *Apollo 15* where he was the Commander. Prior to becoming an astronaut, Scott was an officer in the United States Air Force. Scott may never have predicted the strange experiences he would encounter in his journey to the Moon. In an image of Scott going about his work on the lunar surface where he is seen drilling, an object can be seen hovering above that resembles a spaceship. It appears to be monitoring Scott as he works. One can only imagine what he was thinking if he noticed the object nearby.

In a second incident on the Moon, Scott and astronaut James Irwin were nearly struck by a mysterious object that sped past. What it was that nearly collided with the two men remains unknown to this day. Whatever it was, it was too fast for them to get a good enough look to provide a description. Reportedly, it seemed to have come out of nowhere. This obviously had to have unnerved the two men. What could it have been? Was this a deliberate attempt to startle the astronauts, from someone else on the Moon? And what of the UFO that was nearby as Scott was working? It appears that again, the astronauts were being observed by someone that was perhaps collecting their own data on the astronauts or humans in general. Scott wrote about his trip to the Moon stating, "As I stand out here in the wonders of the unknown at Hadley, I sort of realize there's a fundamental truth to our nature, Man must explore…and this is exploration at its greatest." In an interview, he also commented, "Ever since I was five years old, all I ever wanted to be was a pilot. And flying to the moon seemed to be the ultimate adventure. Nothing seemed more important."

Alan Shepard Jr.

Besides being one of the first twelve men to visit an alien world, Alan Shepard Jr. (1923–1998) has the distinction of being the second person to journey into space. He was the first American to travel into space, which occurred on May 5, 1961. At that time, he served on the Mercury *Freedom 7*. He was also one of NASA's original seven astronauts. In time, he became the Commander of *Apollo 14*, and the fifth person to walk on the Moon.

Shepard once told the *Denver Post*, "I think about the personal accomplishment, but there's more of a sense of the grand achievement by all the people who could put this man on the Moon." On recounting a deeply moving, personal moment on the Moon, Shepard recalled, "When I first looked back on the Earth, standing on the Moon, I cried."

John Young

John Young (1930–2018), as Commander of the *Apollo 16* mission to the Moon, became the ninth person to walk on the surface of the Moon. Prior to becoming an astronaut, he worked as a naval officer, aviator, test pilot and aeronautical engineer. While journeying to the Moon, Young and the other astronauts of *Apollo 16* experienced unexplained flashing lights emitting from an unknown source and reaching their spacecraft. That baffling experience remains a mystery. That withstanding, Young still had hopes for the future of space travel. In a statement made to the Associated Press's journalist Richard Goldstein, and later featured in the article "John Young, Who Led First Space Shuttle Mission, Dies at 87," Young gave his thoughts on the importance of space exploration stating, "Our ability to live and work on other places in the solar system will end up giving us the science and technology that we need to save the species. I'm talking about human beings."

Apollo 1 astronaut Gus Grissom once stated, "If we die, we want people to accept it. We're in a risky business, and we hope that if anything happens to us, it will not delay the program. The conquest of space is worth the risk of life." Grissom's noble words are inspiring, as he was one of the astronauts that perished in the *Apollo 1* accident, in which three lives were lost. In the Moon's galactic history with the Earth, mankind, and any lunar inhabitants, the Apollo program played a great role in connecting it all. All of the sacrifices made for humans to achieve space travel led to the inevitable, which is our return to the cosmos and quite possibly learning that we are indeed part of a galactic community. It was the beginning of something great! *We are looking forward to going into the future!*

In the 1960s NASA's Committee on Long-Range Studies commissioned a study titled the Brookings Report. There is a section in the report titled "The implications of a discovery of extraterrestrial life." This section is believed to have aided NASA in determining how to proceed with the public should extraterrestrial life be discovered. The following are excerpts from the report.

Excerpts from the Brookings Report

(Originally titled: "Proposed Studies on the Implications of Peaceful Space Activities for Human Affairs," by Donald N. Michael)

"While face-to-face meetings with it will not occur within the next twenty years (unless its technology is more advanced than ours, qualifying it to visit earth), artifacts left at some point in time by these life forms might possibly be discovered through our space activities on the Moon, Mars, or Venus." (Pages 182–183)

"Anthropological files contain many examples of societies, sure of their place in the universe, which have disintegrated when they have had to associate with previously unfamiliar societies espousing different ideas and different life ways; others that survived such an experience usually did so by paying the price of changes in values and attitudes and behavior." (Page 183)

252

"Since intelligent life might be discovered at any time via the radio telescope research presently under way, and since the consequences of such a discovery are presently unpredictable because of our limited knowledge of behavior under even an approximation of such dramatic circumstances, two research areas can be recommended:

Continuing studies to determine emotional and intellectual understanding and attitudes—and successive alterations of them if any—regarding the possibility and consequences of discovering intelligent extraterrestrial life.

Historical and empirical studies of the behavior of peoples and their leaders when confronted with dramatic and unfamiliar events or social pressures. Such studies might help to provide programs for meeting and adjusting to the implications of such a discovery. Questions one might wish to answer by such studies would include: How might such information, under what circumstances, be presented to or withheld from the public for what ends? What might be the role of the discovering scientists and other decision makers regarding release of the fact of discovery?" (Pages 183–184)

"An individual's reactions to such a radio contact would in part depend on his cultural, religious, and

SPACE-LIFE REPORT COULD BE SHOCK

The discovery of intelligent space beings could have a severe effect on the public, according to a research report released by the National Aeronautics and Space Administration. The report warned that America should prepare to meet the psychological impact of such a revelation.

The 190-page report was the result of a $96,000 one-year study conducted by the Brookings Institution for NASA's long-range study committee.

Public realization that intelligent beings live on other planets could bring about profound changes, or even the collapse of our civilization, the research report stated.

"Societies sure of their own place have disintegrated when confronted by a superior society," said the NASA report. "Others have survived even though changed. Clearly, the better we can come to understand the factors involved in responding to such crises the better prepared we may be."

Although the research group did not expect any immediate contact with other planet beings, it said that the discovery of intelligent space races "could nevertheless happen at any time."

Even though the UFO problem was not indicated as a reason for the study, it undoubtedly was an important factor. Fear of public reaction to an admission of UFO reality was cited as the main reason for secrecy in the early years of the AF investigation. (Confirmed to NICAP's present director in 1952-3, when the AF was planning to release important UFO reports, also the famous Utah motion-pictures of a UFO formation.)

Radio communication probably would be the first proof of other intelligent life, says the NASA report. It adds: "Evidences of its existence might also be found in artifacts left on the moon or other planets."

This report gives weight to previous thinking by scholars who have suggested that the earth already may be under close scrutiny by advanced space races. In 1958, Prof. Harold D. Lasswell of the Yale Law School stated:

"The implications of the UFOs may be that we are already viewed with suspicion by more advanced civilizations and that our attempts to gain a foothold elsewhere may be rebuffed as a threat to other systems of public order." (UFO Investigator, Dec. 1958.)

The NASA warning of a possible shock to the public, from the revelation of more advanced civilizations, support's NICAP's previous arguments against AF secrecy about UFOs. All available information about UFOs should be given to the public now, so that we will be prepared for any eventuality.

1-11
DEC-JAN 61

253

social background, as well as on the actions of those he considered authorities and leaders, and their behavior, in turn, would in part depend on their cultural, social, and religious environment. The discovery would certainly be front-page news everywhere; the degree of political or social repercussion would probably depend on leadership's interpretation of (1) its own role, (2) threats to that role, and (3) national and personal opportunities to take advantage of the disruption or reinforcement of the attitudes and values of others. Since leadership itself might have great need to gauge the direction and intensity of public attitudes, to strengthen its own morale and for decision making purposes, it would be most advantageous to have more to go on than personal opinions about the opinions of the public and other leadership groups." (Page 183)

"The knowledge that life existed in other parts of the universe might lead to a greater unity of men on earth, based on the 'oneness' of man or on the age-old assumption that any stranger is threatening. Much would depend on what, if anything, was communicated between man and the other beings…" (Page 183)

Annotations from the Brookings Report

"The positions of the major American religious denominations, the Christian sects, and the Eastern religions on the matter of extraterrestrial life need elucidation. Consider the following: 'The Fundamentalist (and anti-science) sects are growing apace around the world …For them, the discovery of other life—rather than any other space product—would be electrifying. …some scattered studies need to be made both in their home centers and churches and their missions, in relation to attitudes about space activities and extraterrestrial life.'" (Page 102)

"If plant life or some subhuman intelligence were found on Mars or Venus, for example, there is on the face of it no good reason to suppose these discoveries, after the original novelty had been exploited to the fullest and worn off, would result in substantial changes in perspectives or philosophy in large parts of the American public, at least any more than, let us say, did the discovery of the coelacanth or the panda." (Page 103)

"If super intelligence is discovered, the results become quite unpredictable. It is possible that if the intelligence of these creatures were sufficiently superior to ours, they would choose to have little if any contact with us. On the face of it, there is no reason to believe that we might learn a great deal from them, especially if their physiology and psychology were substantially different from ours." (Page 103)

"It has been speculated that, of all groups, scientists and engineers might be the most devastated by the discovery of relatively superior creatures, since these professions are most clearly associated with the mastery of nature, rather than with the understanding and expression of man. Advanced understanding of nature might vitiate all our theories at the very least, if not also require a culture and perhaps a brain inaccessible to earth scientists." (Page 103)

"It is perhaps interesting to note that when asked what the consequences of the discovery of superior life would be, an audience of *Saturday Review* readership chose, for the most part, not to answer the question at all, in spite of their detailed answers to many other speculative questions." (Page 103)

"A possible but not completely satisfactory means for making the possibility 'real' for many people would be to confront them with present speculations about the I.Q. of the porpoise and to encourage them to expand on the implications of this situation." (Page 105)

"Such studies would include historical reactions to hoaxes, psychic manifestations, unidentified flying objects, etc. Hadley Cantril's study, *Invasion from Mars* (Princeton University Press, 1940), would provide a useful if limited guide in this area. Fruitful understanding might be gained from a comparative study of factors affecting the responses of primitive societies to exposure to technologically advanced societies. Some thrived, some endured, and some died." (Page 105)

An article from the *UFO Investigator* from 1958 sums up the statements found in the Brookings Report.

"Space-Life Report Could be Shock," *The UFO Investigator Journal* (December 1958)

"The discovery of intelligent space beings could have a severe effect on the public, according to a research report released by the National Aeronautics and Space Administration. The report warned that America should prepare to meet the psychological impact of such a revelation.

The 190-page report was the result of a $96,000 one-year study conducted by the Brookings Institution for NASA's long-range study committee.

Public realization that intelligent beings live on other planets could bring about profound changes, or even the collapse of our civilization, the research report stated.

"Societies sure of their own place have disintegrated when confronted by a superior society," said the NASA report. "Others have survived even though changed. Clearly, the better we can come to understand the factors involved in responding to such crises the better prepared we may be."

Although the research group did not expect any immediate contact with other planet beings, it said that the discovery of intelligent space races "could nevertheless happen at any time."

Even though the UFO problem was not indicated as a reason for the study, it undoubtedly was an important factor. Fear of public reaction to an admission of UFO reality was cited as the main reason for secrecy in the early years of the AF investigation. (Confirmed to NICAP's present director in 1952-53, when the AF was planning to release important UFO reports, also the famous

President John F. Kennedy giving his Moon Speech at Rice Stadium, 1962

Utah motion-pictures of a UFO formation.)

Radio communication probably would be the first proof of other intelligent life, says the NASA report. It adds: "Evidences of its existence might also be found in artifacts left on the moon or other planets."

On September 12, 1962, President John F. Kennedy gave a speech at Rice Stadium. It was the President's intent to convince the public to support the Apollo program. Some refer to the speech as the "We choose to go to the Moon" speech. It is also referred to at the "Address at Rice University on the Nation's Space Effort."

President John F. Kennedy's Rice Stadium Moon Speech

President Pitzer, Mr. Vice President, Governor, Congressman Thomas, Senator Wiley, and Congressman Miller, Mr. Webb, Mr. Bell, scientists, distinguished guests, and ladies and gentlemen:

I appreciate your president having made me an honorary visiting professor, and I will assure you that my first lecture will be very brief.

I am delighted to be here, and I'm particularly delighted to be here on this occasion.

We meet at a college noted for knowledge, in a city noted for progress, in a state noted for strength, and we stand in need of all three, for we meet in an hour of change and challenge, in a decade of hope and fear, in an age of both knowledge and ignorance. The greater our knowledge increases, the greater our ignorance unfolds.

Despite the striking fact that most of the scientists that the world has ever known are alive and working today, despite the fact that this Nation's own scientific manpower is doubling every 12 years in a rate of growth more than three times that of our population as a whole, despite that, the vast stretches of the unknown and the unanswered and the unfinished still far outstrip our collective comprehension.

No man can fully grasp how far and how fast we have come, but condense, if you will, the 50,000 years of man's recorded history in a time span of but a half-century. Stated in these terms, we know very little about the first 40 years, except at the end of them advanced man had learned to use the skins of animals to cover them. Then about 10 years ago, under this standard, man emerged from his caves to construct other kinds of shelter. Only five years ago man learned to write and use a cart with wheels. Christianity began less than two years ago. The printing press came this year, and then less than two months ago, during this whole 50-year span of human history, the steam engine provided a new source of power.

Newton explored the meaning of gravity. Last month electric lights and telephones and automobiles and airplanes became available. Only last week did we develop penicillin and television and nuclear power, and now if America's new spacecraft succeeds in reaching Venus, we will have literally reached the stars before midnight tonight.

This is a breathtaking pace, and such a pace cannot help but create new ills as it dispels old, new ignorance, new problems, new dangers. Surely the opening vistas of space promise high costs and hardships, as well as high reward.

So it is not surprising that some would have us stay where we are a little longer to rest, to wait. But this city of Houston, this State of Texas, this country of the United States was not built by those who waited and rested and wished to look behind them. This country was conquered by those who moved forward—and so will space.

William Bradford, speaking in 1630 of the founding of the Plymouth Bay Colony, said that all great and honorable actions are accompanied with great difficulties, and both must be enterprised and overcome with answerable courage.

If this capsule history of our progress teaches us anything, it is that man, in his quest for knowledge and progress, is determined and cannot be deterred. The exploration of space will go ahead, whether we join in it or not, and it is one of the great adventures of all time, and no nation which expects to be the leader of other nations can expect to stay behind in the race for space.

Those who came before us made certain that this country rode the first waves of the industrial revolution, the first waves of modern invention, and the first wave of nuclear power, and this generation does not intend to founder in the backwash of the coming age of space. We mean to be a part of it—we mean to lead it. For the eyes of the world now look into space, to the moon and to the planets beyond, and we have vowed that we shall not see it governed by a hostile flag of conquest, but by a banner of freedom and peace. We have vowed that we shall not see space filled with weapons of mass destruction, but with instruments of knowledge and understanding.

Yet the vows of this Nation can only be fulfilled if we in this Nation are first, and, therefore, we intend to be first. In short, our leadership in science and in industry, our hopes for peace and security, our obligations to ourselves as well as others, all require us to make this effort, to solve these mysteries, to solve them for the good of all men, and to become the world's leading space-faring nation.

We set sail on this new sea because there is new knowledge to be gained, and new rights to be won, and they must be won and used for the progress of all people. For space science, like nuclear science and all technology, has no conscience of its own. Whether it will become a force for good or ill depends on man, and only if the United States occupies a position of pre-eminence can we help decide whether this new ocean will be a sea of peace or a new terrifying theater of war. I do not say the we should or will go unprotected against the hostile misuse of space any more than we go unprotected against the hostile use of land or sea, but I do say that space can be explored and mastered without feeding the fires of war, without repeating the mistakes that man has made in extending his writ around this globe of ours.

There is no strife, no prejudice, no national conflict in outer space as yet. Its hazards are hostile to us all. Its conquest deserves the best of all mankind, and its opportunity for peaceful cooperation many never come again. But why, some say, the moon? Why choose this as our goal? And they may well ask why climb the highest mountain? Why, 35 years ago, fly the Atlantic? Why does Rice play Texas?

We choose to go to the moon. We choose to go to the moon in this decade and do the other things, not because they are easy, but because they are hard, because that goal will serve to organize and measure the best of our energies and skills, because that challenge is one that we are willing to accept, one we are unwilling to postpone, and one which we intend to win, and the others, too.

It is for these reasons that I regard the decision last year to shift our efforts in space from low to high gear as among the most important decisions that will be made during my incumbency in the office of the Presidency.

In the last 24 hours we have seen facilities now being created for the greatest and most complex exploration in man's history. We have felt the ground shake and the air shattered by the testing of a Saturn C-1 booster rocket, many times as powerful as the Atlas which launched John Glenn, generating power equivalent to 10,000 automobiles with their accelerators on the floor. We have seen the site where the F-1 rocket engines, each one as powerful as all eight engines of the Saturn combined, will be clustered together to make the advanced Saturn missile, assembled in a new building to be built at Cape Canaveral as tall as a 48 story structure, as wide as a city block, and as long as two lengths of this field.

Within these last 19 months at least 45 satellites have circled the earth. Some 40 of them were "made in the United States of America" and they were far more sophisticated and supplied far more knowledge to the people of the world than those of the Soviet Union.

The Mariner spacecraft now on its way to Venus is the most intricate instrument in the history of space science. The accuracy of that shot is comparable to firing a missile from Cape Canaveral and dropping it in this stadium between the 40-yard lines.

Transit satellites are helping our ships at sea to steer a safer course. Tiros satellites have given us unprecedented warnings of hurricanes and storms, and will do the same for forest fires and icebergs.

We have had our failures, but so have others, even if they do not admit them. And they may be less public.

To be sure, we are behind, and will be behind for some time in manned flight. But we do not intend to stay behind, and in this decade, we shall make up and move ahead.

The growth of our science and education will be enriched by new knowledge of our universe and environment, by new techniques of learning and mapping and observation, by new tools and computers for industry, medicine, the home as well as the school. Technical institutions, such as Rice, will reap the harvest of these gains.

And finally, the space effort itself, while still in its infancy, has already created a great number of new companies, and tens of thousands of new jobs. Space and related industries are generating new demands in investment and skilled personnel, and this city and this State, and this region, will share greatly in this growth. What was once the furthest outpost on the old frontier of the West will be the furthest outpost on the new frontier of science and space. Houston, your City of Houston, with its Manned Spacecraft Center, will become the heart of a large scientific and engineering community. During the next 5 years the National Aeronautics and Space Administration expects to double the number of scientists and engineers in this area, to increase its outlays for salaries and expenses to $60 million a year; to invest some $200 million in plant and laboratory facilities; and to direct or contract for new space efforts over $1 billion from this Center in this City.

To be sure, all this costs us all a good deal of money. This year's space budget is three times what it was in January 1961, and it is greater than the space budget of the previous eight years combined. That budget now stands at $5,400 million a year—a staggering sum, though somewhat less than we pay for cigarettes and cigars every year. Space expenditures will soon rise some more, from 40 cents per person per week to more than 50 cents a week for every man, woman and child in the United Stated, for we have given this program a high national priority—even though I realize that this is in some measure an act of faith and vision, for we do not now know what benefits await us.

But if I were to say, my fellow citizens, that we shall send to the moon, 240,000 miles away from the control station in Houston, a giant rocket more than 300 feet tall, the length of this football field, made of new metal alloys, some of which have not yet been invented, capable of standing heat and stresses several times more than have ever been experienced, fitted together with a precision better than the finest watch, carrying all the equipment needed for propulsion, guidance, control, communications, food and survival, on an untried mission, to an unknown celestial body, and then return it safely to earth, re-entering the atmosphere at speeds of over 25,000 miles per hour, causing heat about half that of the temperature of the sun—almost as hot as it is here today—and do all this, and do it right, and do it first before this decade is out—then we must be bold.

I'm the one who is doing all the work, so we just want you to stay cool for a minute. [laughter]

However, I think we're going to do it, and I think that we must pay what needs to be paid. I don't think we ought to waste any money, but I think we ought to do the job. And this will be done in the decade of the sixties. It may be done while some of you are still here at school at this college and university. It will be done during the term of office of some of the people who sit here on this platform. But it will be done. And it will be done before the end of this decade.

I am delighted that this university is playing a part in putting a man on the moon as part of a great national effort of the United States of America.

Many years ago, the great British explorer George Mallory, who was to die on Mount Everest, was asked why did he want to climb it. He said, "Because it is there."

Well, space is there, and we're going to climb it, and the moon and the planets are there, and new hopes for knowledge and peace are there. And, therefore, as we set sail, we ask God's blessing on the most hazardous and dangerous and greatest adventure on which man has ever embarked.

Thank you.

Chapter Ten

Into the Future

In my own view, the important achievement of Apollo
was a demonstration that humanity is not forever chained
to this planet, and our visions go rather further than that,
and our opportunities are unlimited.
—Neil Armstrong

When I started writing this book, I remembered a movie that I had watched as a child. Seen through a young person's eyes, it left an impression on me. It was the film comedy *Way Way Out!* It stared Jerry Lewis and Connie Stevens, two of my favorites! It was a silly little comedy that took place on a minute settlement on the Moon, and *I loved it*! Even though the movie graphics were nowhere as sensational as what we have today, the futuristic imagery (for that time) was memorable. I recently rewatched that movie just to get a view once again of their lunar station. Although I still enjoyed it and felt a sense of nostalgia watching it, I realized that it is nothing near what my vision of a lunar base would be today, nor my vision of man's future in space in general.

When we look towards the future I ask, will we, to quote *Star Trek*, *"boldly go where no person has gone before?"* Will we take the next step in our journey to the stars? If so, when, and just how far will we go? The words of Neil Armstrong once again come to mind. "There are great ideas undiscovered, breakthroughs available to those who can remove one of truth's protective layers." Armstrong's words were prophetic to what is going on today. We are trying to remove the "protective layers" of what has been hidden from us for so long and answer the questions of the secrets of the universe that have been illusive to mankind. We have reached a pinnacle in our thinking about our past, how it

Advertisement for the movie *Way Way, Out* starring Jerry Lewis
and Connie Stevens (1966).

connects to the universe and how it will affect our future. We are more open than ever before to finding answers to the questions that have plagued mankind since the beginning. Questions such as, *"Who are we?" Why are we here? What is our purpose? How did we get here? Are we alone in the universe?* These questions are slowly being answered. Truths are being uncovered. New mysteries are being discovered. All questions cannot and will not be answered all at once. It is a painstakingly slow process to find the answers. We may have to reach out into the stars to receive the answers to most of our questions. Many of the answers, we will not know until we get out there. Who knows what the future will bring and what we will encounter?

This reminds me of a conversation from the original classic movie *Planet of the Apes*. This movie portrayed a time in which apes and monkeys were the intelligent species on Earth and humans were regarded as animals. Somewhere in time, the roles had become reversed on Earth. Taylor, an astronaut who had been absent for a very long time, was shocked to discover such a society. Eventually, Taylor was caught and locked up. Through a series of dramatic, heart-wrenching and thought-provoking events, the authorities unwillingly released him. He was free to follow his own path. In the end, Taylor was leaving the city, traveling by horseback with his mate Nova. Taylor and Nova together were going to face their uncertain future. As the apes watched them leave, headed toward what the apes called "the forbidden zone," the chimpanzee psychologist Zira asks Zaius (the ape in charge of society) a compelling question. Zira asks, "What will he find out there, doctor?" Zaius answers, "His destiny." As Taylor journeys on his path, he finds the Statue of Liberty practically buried in sand, with only the upper body, the head and the arm holding the torch protruding upward. He then realizes that mankind had destroyed themselves and the world he once knew. Taylor had found his answers and his destiny. Will we find the answers to our past in the stars? *Are the cosmos our destiny?*

Once we return to the Moon and journey to Mars (and hopefully beyond), we just may learn the truth about our past, and quite possibly discover that we are a part of a universe that is teeming with life. If mankind were to learn that we are not alone, then we would be propelled into a reality that is far more fascinating than anything humans have ever known. I believe too that we are slowly being introduced to the idea that we are a part of a universal community. We see it in our media, our books, television, movies, social media and more; and the truths that are being uncovered and revealed are astonishing! It almost seems that we are being quietly prepared for the disclosure that we are not alone. So then, what is the future of visiting the Moon and space travel in general? Let's look at the possibilities.

NASA

It has been fifty years since the last Apollo mission to the Moon. Since that time there has been plenty of speculation as to why we have not returned, with budget cuts and loss of interest on the public's behalf topping the list. Today, NASA is stating that it will be sending people to the Moon in the near future. The goal as of this writing is the Moon's South Pole. NASA's website states, "NASA is implementing the President's Space Policy Directive-1 to 'lead an innovative and sustainable program of exploration with commercial and international partners to enable human expansion across the solar system.'" If all goes as planned, then NASA will be returning astronauts to the Moon soon. Perhaps then we will find the answers to the many mysteries involving the Moon. Of course, there is still the possibility of a cover-up of some of the answers about the Moon that are still being sought. Will we find proof of an ancient past connection to the Moon? Will we be told if it is discovered that ancient ruins on the Moon are connected to Earth? Will we know if there are extraterrestrials or humans residing on the Moon? *Will they tell us?*

Why Return to the Moon?

Most people consider the Moon to be an inhospitable place to visit, let alone build a colony. It has extremely heated daytime temperatures and freezing nights. There is also unbridled cosmic radiation and weak gravity. However, even with the risks, it is beginning to look as though it may be a necessity to have a colony in space, with the Moon being one of the main candidates for a place to do so. Currently (and this could change) the proposed site is located on the Moon's South Pole where there is continuous daylight. Some wonder why we would need to set up a base on the Moon, since it is so inhospitable? One of the reasons we should send manned missions to the Moon is what I refer to as "the great escape." The Earth is volatile. It is subject to disasters of all sorts that could wipe out our species. Although it is not something that most people consider as they go about their daily lives, there have been several disasters in Earth's past that have devastated civilizations. It could happen again. There is the possibility that mankind has had to start over several times in our advancement, due to world catastrophes.

Events that could present a situation on Earth from which people have to escape could consist of anything from a massive asteroid or meteor impact to an atomic war, a global pandemic, an environmental disaster, a super volcano, another ice age, or the planet overheating due to climate change. A cataclysmic event could happen on Earth at any time and life would be wiped out. There is also the dwindling of resources on Earth. We may find ourselves in a position sometime in the future where humankind needs to escape Earth to preserve the species. A colony off planet is our best option for survival. For now, the Moon seems to be the most logical place to start given its proximity. If not used for settlements, it could be used as a launch point to establish a colony on Mars. We should go into the future with the idea of being a spacefaring civilization, starting with new missions to the Moon. Of course, given what we know about possible life on the Moon, we may need permission. If that cannot be achieved, then it is on to Mars, which ironically enough, may have the same situation (but that is another book).

Looking Forward...

Some of us are dreamers. To escape the mundane of everyday living, we escape into movies, literature, and other media as a means of stepping out of our own lives and into something incredible, *something more*. That is when fantasy comes in. That is where science fiction begins. When I consider the future, I dream of a time when we will connect with a galactic community. I also look forward to the time of first contact. This would be the introduction of off-world beings that have our best interests at heart and would accept friendship based on peace. I think about what could be, as we begin our sojourn into space and begin to establish colonies on other worlds. I dream of space exploration, the wonders of it, and the places it will take us. I consider where we might go as citizens of the universe, where we may travel to and ultimately what other intelligent lifeforms we will encounter. So just what are we considering today that will propel mankind into the future and such an amazing new reality? There is talk of NASA preparing mankind for disclosure and possibly first contact. There is also talk of creating habitats on various cosmic bodies throughout the solar system beyond the Moon and Mars. Beyond that who knows what may happen. *The sky is literally the limit!*

First Contact

Are we slowly being prepared for the day when extraterrestrials will introduce themselves to us? There are those who believe that we are subtly being prepared for disclosure through the various forms of media. Others maintain that we will not reach the disclosure that we are not alone until we become a Type 1 civilization. On the Kardashev Scale a Type 1 civilization can harness all the energy that falls on a planet from its parent star. It is the technological level that is near to what we presently have achieved on Earth. It is thought that we will reach Type 1 status in 100-200 years. From a realistic standpoint, humans are a very young species.

Other worlds have had eons of time to grow and advance, whereas we are just getting started in the area of space travel. There could be extraterrestrial civilizations that are millions if not billions of years ahead of us. This is one reason why some believe the Moon may have been brought here. There are beings that may be so far advanced, that to build a gigantic, camouflaged ship and navigate it across the universe may have been doable. Is this not what we as humans seek to achieve? Do we not want to grow? Do we not want to reach the stars and place colonies on other worlds? Extraterrestrials from other worlds may already have achieved what we are just beginning to do.

In our search for intelligent life in the cosmos, the Moon is an excellent place to start. In fact, it looks as though we did not travel very far from Earth before we ran into someone out there. If just one of the stories of the experiences of the astronauts in space seeing other spacecraft, or intelligently-made structures on the Moon, or being followed to the Moon, and more, then that means that we are not alone in the universe. For those who believe we never went to the Moon, look at the evidence. Look at the images from the Apollo missions. Listen to the testimonies of NASA astronauts, officials and employees and others that have come forward to try to educate people about what is out there. As former astronaut Edgar Mitchell once stated, "Read the books, read the lore, start to understand what has really been going on, because there is no doubt that we are being visited. The universe that we live in is much more wondrous, exciting, complex, and far-reaching than we were ever able to know up to this point in time. Mankind has long wondered if we're alone in the universe. But only in our period do we really have evidence. No, we're not alone."

There is a story that the creator of *Star Trek*, Gene Roddenberry, was sitting in on channeling sessions with advanced extraterrestrials. Some believe that Mr. Roddenberry received his idea for the *Star Trek* series from those sessions. The beings that were channeling the information reportedly stated that they were in Earth's vicinity on a huge ship. They stated that the ship held many beings from different worlds, and that they were working together in unison. They stated that they were peacefully traveling the universe exploring and doing good works.

If you look at some of the amazing episodes of the original *Star Trek* series, *Star Trek: The Next Generation* and the others, there are very interesting themes. The shows are said to be inspired by the travels of these beings, with entire manuscripts being written around who they are and what they stood for. Some of the most interesting shows included such stories as, a creator seeding planets, gods that had retired due to the loss of interest in them from humans, extraterrestrials that were of a hive mind, hybrid children of alien and human unions, shapeshifters, time travel, a civilization existing under the surface of an asteroid (see Chapter Two), traveling through portals and stargates, warp drive and more.

A photo of *Star Trek* creator Gene Roddenberry and the original cast at NASA.

There is so much we are hearing about today in our own research and understanding of UFOs, extraterrestrials, time, space and other related topics. In the *Star Trek* universe, humans were the heroes with a purpose, ready to assist and find their place in the universe. Will that be *us*? Or will we be the planet in an apocalyptic ruinous state as seen in so much of the science fiction of today? Predictions fortunately are not written in stone. We can shape our future when it comes to space travel and space exploration. We should look to traveling the stars not only to grow as a species, but in case Earth faces that devastating, cataclysmic scenario that I mentioned earlier. We would have someplace else to go. In such planning, we would be known as the generation that looked ahead and prepared. That is what our future should be.

Back to the Moon...

To some, the idea that the Moon may be inhabited remains a crazy thought! But I ask you, what is wrong with exploring the possibility? What is wrong with furthering the notion that we are not alone in the universe? How will we ever progress and advance

270

into the cosmos if we do not investigate and explore the closest planetary body to us? At the least, we will have gained answers to the many mysteries surrounding our Moon. Perhaps it has an existential relationship to humans. The argument is that we have been to the Moon, sent probes to the Moon and no one has been found. The truth is we have explored less than one percent of the lunar surface. We have not explored by any means all of the Moon. If that is the biggest argument for there not being intelligent life on the Moon after all that we have covered here, there is not much of a case there. However, when we examine the evidence which includes images of intelligently-made structures, testimony from astronauts, recordings of conversations between astronauts, comments from other countries that have sent up lunar probes divulging odd incidents in their missions, testimonies from people that worked for NASA, records of strange lights, movement on the Moon and more, then I would say that there is a very good case to warrant further investigation of the Moon, and determine for certain whether or not there is intelligent life there. Most importantly, if we are able, *we should find out who they are!* The following is what we have to look forward to as we proceed into the future:

- Disclosure that we are not alone in the universe
- The establishment of a Moon colony (or at the very least a lunar base)
- A colony on Mars
- Warp speed
- Space Exploration
- Settlements throughout the galaxy

NASA has yet to come forward about what was found on the Moon. I mentioned before that that is probably due to the Brookins Report. However, considering the importance of the Moon, and the fact that life on Earth relies on it, it would seem that the Moon would be at the forefront of attention. People should not be left to their own devices to figure out what is going on with the Moon. If something happens to the Moon, much (if not all) of life on Earth would be wiped out. Because of this, there should be more studies of the Moon and more understanding of its origin. There should

271

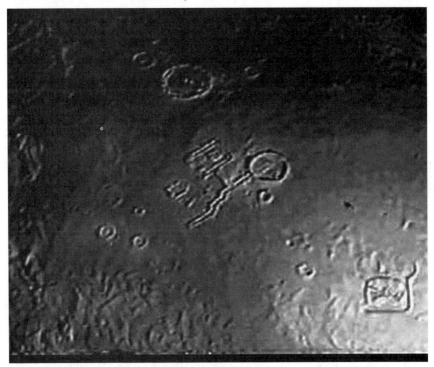

NASA photo of a Moon city in the UK *Daily Star* on April 13, 2018.

be conversations about what we would do if something happened to the Moon. *(Seriously!)* Also, if there are beings residing there, we need to take it seriously and understand who or what we are dealing with. Are they visiting Earth? Are they responsible for the UFOs seen around the globe? Are they involved in alien abductions, cattle mutilations, and the disappearance of water? Are they using our resources? Are they sending us messages? Are they responsible for crop circles? If they are there, is it possible to establish a peaceful relationship, friendship or even an alliance with them? If they are in the position to move the Moon away from Earth, that is problematic. This may all sound like a fantastic science fiction movie. However, there *is* something going on with the Moon, and it would behoove us to learn what.

On November 23, 2021, it was officially announced that an office was opened in the Pentagon titled the Airborne Object Identification and Management Synchronization Group (AOIMSG). It is tasked with investigating UFOs (UAPs) that enter US airspace. Many feel that with the establishment of this office, progress has been made in the national interest in the UFO subject. With this office there are those who are hopeful that we are moving toward disclosure, which is the opening of records of what the government knows about our relationship with extraterrestrials, as well as information about UFOs. As a result, many hope that disclosure is not far off. Others contend that disclosure will lead to first contact. Possibly. However, as I indicated before, it may not be the governments that believe we are not ready for disclosure, it may well be the extraterrestrials that feel that we are not ready, and this may be the reason we have had no "official" word about them. If true, one would imagine that there just may be a timeline made by the extraterrestrials that are watching our advancement as to when we will be prepared for this news.

What the Future May Hold

In our move toward the lofty goal of space exploration, there are things we should keep in mind as we contemplate our future. First and foremost, we need return to the Moon. Once we do, it would be prudent to:

1. Locate tangible evidence of the Moon being inhabited.
2. Gather proof that there was once an ancient civilization on the Moon.
3. Offer complete transparency to the public of the discoveries on the Moon.
4. Establish a lunar base.

What is very exciting is that there is already talk about colonizing various areas of the solar system. Some promising sites have already been considered for habitation. None of these are in the planning stages. They are only ideas that have been looked at by scientists. However, it is exhilarating to know that our future in space is being seriously thought about and is in the early planning stages. Some of the areas that have been considered for human habitation include the Moon, Mars,

Pyramid on the Moon from Pinterest.

273

Venus, asteroids, Jupiter's moon Europa, the Bernal Sphere, Mercury and the methane lakes of Titan. Please note that except for the Moon and Mars, which are discussed seriously, these are only ideas. Perhaps one day they will come to fruition. But for now, they make for very interesting reading and contemplation.

Space Colonies of the Future

Asteroids

In an interview with *Mother Earth News, Science, Technology and Space Magazine*, sci-fi great Isaac Asimov gave his views on using asteroids for colonization, stating, "Asteroids would—by the way—make for much better settlements than would the moon, because that satellite doesn't have certain lightweight ele-

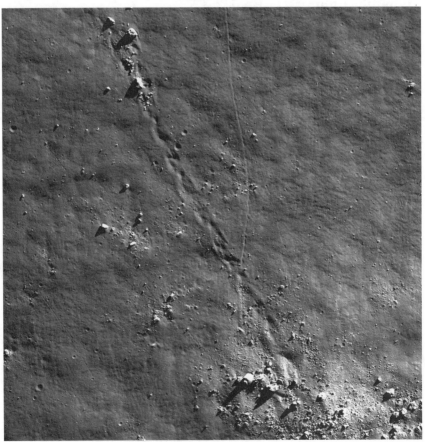

Photo of a moving object taken by the Lunar Reconnaissance Orbiter Camera.

ments... including carbon, hydrogen, and nitrogen. Lunar residents would therefore have to depend on the Earth for such basic substances, but settlers of the asteroids could have their own supplies and be truly independent of Earth." Even though most

Two craters with "Moon cities" from Pinterest.

are looking forward to a time when we may possibly colonize the Moon, Mars, and other planetary objects, the idea of creating a colony on asteroids is being seriously considered.

Asteroids are thought by many to simply be large boulders drifting in space with metals inside of them. Some of the metals on these asteroids are thought to be valuable and could most likely be utilized back home on Earth (although the cost to transport them may nullify the actual profit). However, asteroids can play an important role in the survival of mankind in case of a worldwide catastrophe. To date 300,000 asteroids have been identified.

Therefore, the colonization of asteroids could assist in humans spreading themselves around the solar system. Additionally, they contain various chemical composition classes, such as ferrous and carbonaceous, providing a variety of substances usable in constructing as well as powering spacecraft, bases, and colonies. These asteroid colonies would most likely be located beneath the surface of the asteroid. I might add here that there is a possibility that this may already be the case. We read in Chapter Two about the hollow Moon and a nod was given to a *Star Trek* episode about a city within an asteroid. If the Moon is hollow and there are beings with a habitat inside, who'se to say that this has not already occurred with asteroids somewhere in our solar system?

The Moon

There was talk of establishing a base on the Moon even before the astronauts landed on the lunar surface. Now, it has been over five decades since that tremendous day when humans first walked on the Moon. Today, both governments and private corporations are devising plans to put bases on the Moon, with the hope that they may eventually develop into a full-blown colony, as well as provides a jump off point for travel into space beginning with Mars. It will be however, a tremendous challenge, as the conditions on the lunar surface are harsh. The Moon's rotation cycle

A photo of China's Chang'e-4 from its rover.

is 28 days. Therefore, there are two weeks of daytime light and two weeks of nighttime darkness.

The days are sweltering. Temperatures can reach up to 130 degrees. The nights are freezing, with temperatures plummeting to minus 160 degrees. The cosmic radiation coming from the universe is radioactive. There is also the problem of the Moon dust. Moon dust is very fine. The astronauts found it difficult to work in. It has been compared to the consistency of volcanic ash, and tends to make its way into machinery, sticks to spacesuits, etc. Companies must focus on how to basically keep people alive on the Moon. They must consider food, power, water, oxygen, and other necessities. The bases will have to be pre-made and transported to the Moon from Earth, so they can be utilized right away.

Some companies have already created prototype lunar habitats, including an expandable portable habitat. The habitat would be transported aboard a rocket. In contemplating the idea of an installation there, some scientists have considered a transportable base on wheels, that can be moved according to where the sunlight is heating, during the long cold lunar nights. With that said, a base would likely consist of a landing port, a science area, housing, a greenhouse, water utilities, a mining facility, a mess hall, an observatory, medical quarters and possibly a recreation hall.

One idea that is being considered is for private citizens to travel to the Moon for recreation. Although, this would be secondary to a base and a lunar colony, the Moon may one day be used as a place for people to travel to for the experience of journeying to outer space. We may even find space hotels set up for people to visit the Moon and take trips around the lunar surface. There may even be a lunar observatory for those interested in astronomy.

A photo of the lunar rover from the Chang'e-4.

A Space Elevator

For future space travel, the idea of a space elevator has been entertained as a means of transportation into space. The concept was first introduced to the public by Arthur C. Clarke in his novel titled *The Fountains of Paradise* (1979). A space elevator is a type of Earth-to-space conveyance system. It would allow humans to travel into space more efficiently and cheaply than large rockets. The most important feature would be a tether that would be anchored to Earth's surface and extend out into space. Vehicles would travel alongside the tether, and into orbit. This gigantic elevator would take people from Earth into space with the Moon and Mars as destinations, and in the future, perhaps even further. The idea is on hold for now, as currently there is no known material strong enough to create this device. However, scientists are working on it, hoping that it will be a viable plan in the future.

The Bernal Sphere

The Bernal Sphere is a concept for an independent, self-sufficient, orbital galactic space station. It was originally proposed in 1929 by Physicist John Desmond Bernal (1901–1971). Bernal was known for his research into the atomic structure of solid compounds. He played a key role in x-ray crystallography. Bernal's idea for this megastructure has been compared to that of the science fiction series *Babylon 5*. The sphere is basically an air-filled sphere, spun up to supply gravity to its residents near the equator. The initial plan was of a sphere measuring nearly 10 miles in diameter. This would provide enough area for a populace of approximately 20,000 to 30,000 to live and work comfortably in space. The station would include all the essential items needed

277

for survival and more. It would be encircled by gigantic rings that are continuously spinning at high velocities. Each ring would be designed with a distinctive function in mind. For example, one would be used for producing crops and another for the upkeep of livestock. The station would also have Earthlike gravity due to its constant rotating which would occur twice per minute.

Europa

Incredibly, Europa, one of the moons of Jupiter, has been considered for a future space settlement. It is thought that it just may be a suitable place to establish a colony. This is extremely interesting because Europa is enclosed in a heavy sheet of ice. However, scientists suspect that under the ice is an entire ocean. The exterior of Europa is overrun with radiation from both the Sun and Jupiter. However, beneath the ice, people would be protected from the radiation. Theoretically, there should be areas with compartments filled with air dividing the water from the ice. It is thought that in those spaces, floating residences could be established. It is also believed that we could possibly fill the belowground seas of Europa with a variety of aquatic life!

One other interesting aspect of Europa is that there has been talk of intelligent life possibly being located there. Not only might we have a colony there one day, but there may be extraterrestrial neighbors there as well. In a 2012 article for *Live Science*, titled "If We Discover Aliens, What's Our protocol for Making Contact?" by Natalie Wolchover, Astronomer Jacob Haqq-Misra, of Pennsylvania State University is quoted as saying, "It is consistent with current human exploration of the solar system that intelligent beings could have evolved in the deep oceans of Europa." Time will tell, but it appears that Europa is an exciting prospect all around.

Mars

As most know, there are plans to colonize Mars in the works. Mars is definitely a goal when it comes to future space travel and creating a settlement. Mars is for now, the main backup plan if a calamitous event were to threaten life on Earth. The plan being considered is a habitat with glass domes on the surface. This city is thought to be on track to happen in 50 to 100 years. If it reaches fruition, then many alive today, will live to see it. What a thrilling prospect for the future!

Mercury

Domed environments on Mercury have been considered as an appropriate place for human habitation as well. A mobile city that is self-sustaining, with a breathable atmosphere could be created, even though it is the nearest planet to the sun! In another proposed idea, all human habitation and agriculture would be established below the surface to avoid Mercury's extreme temperatures, ionizing radiation, and other perilous elements to humans.

Titan

The American Aerospace Engineer and Author Robert Zubrin once stated, "In certain ways, Titan is the most hospitable extraterrestrial world within our solar system for human colonization." Titan, one of Saturn's moons, has been referred to as "a very strange place." Methane pours from the skies which eventually creates large lakes. The idea of colonizing this place sounds unbelievable. Still, it is thought to be a good place to for human habitation. The reason is that one of the greatest dangers to existing in space is the radiation from the galactic cosmic rays. In fact, the best place to escape these cosmic rays in the entire solar system is Titan. The atmosphere there consists of quite a bit of nitrogen which is helpful in impeding cosmic rays. Titan also has heightened shielding from Saturn's magnetosphere. Therefore, if we were able to discover a way to live there, it would in fact be one of the most innocuous areas in space for a settlement. A benefit of living on Titan is that the lakes of methane and ethane are great energy and power resources. This energy could be utilized in the colonies which would be made from plastic domes that would be inflated by warm nitrogen and oxygen. Earth would also benefit from this as the fuels could be used here as well.

Venus

Venus is known as Earth's double. It is said to be the planet with an environment that is most similar to that of Earth, and therefore is a good place for a colony. However, it is in the *clouds* of Venus that temperatures are similar to Earth's. As a result, it is in the clouds that floating cities have been considered as a place for the future colonization of humans. This city would literally be held up with large balloons in the safe atmospheric area in Venus' skies. It is one of the few places that could hold a human settlement if things ever went awry on Earth.

Cosmic Mysteries to Explore

As we head into the future of space travel, we have no idea what we will discover out there. Interestingly, we already have many cosmic mysteries to solve. We are working on the mysteries of the Moon, and one day, hopefully soon, we will find the answers. Mars too is filled with secrets to explore. UFOs have been seen in the skies above, and anomalous objects have been located on the surface. There is also a theory that Mars' moons Phobos and Deimos are hollow as well. Additionally, as stated before, the Earth, Moon and Mars may have been engaged in war at one time. One can only wonder, if this is true, were we all fighting against each other, *or were we fighting an unknown foe?*

As we move into the future, we may be investigating such cosmic mysteries as those of Oumuamua and the Black Knight Satellite. Oumuamua is an enormous, unexplained cylinder-shaped object seen traveling through our solar system. Some scientists believe that it may be intelligently created. Due to the uncommon characteristics of this visitor, it is believed to have come from outside of our solar system. It is thought to be an alien probe, or even a spaceship! Additionally, there is the Black Knight Satellite, a mysterious object that has been observed for years, hovering above our atmosphere. It has also been observed near the Moon. There are those who believe it to possibly be an alien probe sent from outside of our universe. There is also the mystery of the monolith that sits on Mars' moon Phobos. Who put it there? What is its purpose? There are many other mysteries happening out in space that we do not have the answers to. Therefore, who knows what awaits us in space and in our future. *There are many cosmic mysteries to explore as we go into the future!*

Author Biography

Constance Victoria Briggs is an author, researcher, and guest speaker on cosmic and metaphysical topics. She specializes in Moon mysteries and phenomena. She is the author of *The Encyclopedia of Angels: An A-to-Z Guide with Nearly 4,000 Entries; Encyclopedia of God: An A-Z Guide to Thoughts, Ideas, and Beliefs about God; Encyclopedia of the Unseen World: The Ultimate Guide to Apparitions, Death Bed Visions, Mediums, Shadow People, Wandering Spirits and Much, Much, More,* and *The Encyclopedia of Moon Mysteries: Secrets, Conspiracy Theories, Anomalies, Extraterrestrials and More.* Briggs is a regular on radio shows, podcasts and YouTube shows discussing such topics as angels, extraterrestrials, ancient astronauts, life-after-death, after-death communication, out-of-body experiences, Moon mysteries and more. Shows that Briggs has been featured on include *Coast to Coast am with George Noory, Midnight Society, The Leak Project, The Kingdom of Nye, Earth Ancients, Paranormal Soup, The Paranormal and the Sacred, Broadcast Team Alpha, Forbidden Knowledge, Quantum Guides, Behind the Obsidian Curtain, Spaced Out Radio, Raised by Giants,* and many others. Briggs has also been featured in the popular, *Shadows of Your Mind Magazine.* Constance is known as the "Moon Mysteries Expert," and was recently a guest speaker at the popular *Total Disclosure: Inner Truth Summit.* It is Briggs's goal to investigate the mysteries of the universe and how they connect to humanity. States Constance, "There is more to our universe than we have been taught. We need to look deeper and search further if we want to know the truth of who we are, where we come from, and what lies beyond."

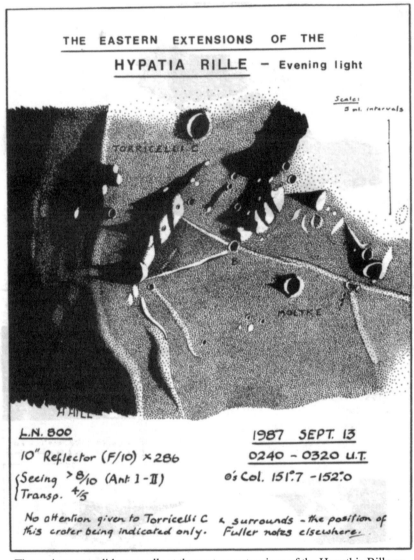

The curious monoliths or walls at the eastern extensions of the Hypathia Rille as drawn by the British Astronomer Howard Hill.

Bibliography

Book Sources

Above Top Secret: UFO's, Aliens, 9/11, NWO, Police State, Conspiracies, Cover Ups, and Much More "They" Don't Want You to Know About. Jim Marrs. The Disinformation Company Ltd. New York, NY. 2008.

Alien Threat from the Moon: Evidence of Ancient and Alien Life. Dylan Clearfield. G. Stempien Publishing Company. Blessed Isle, Wales, UK. 2016.

Aliens on the Moon: The Moon is the Key to the Secrets of the Aliens on Earth! Gil Carlson, Blue Planet Press, Hampton VA. 2017.

Aliens on the Moon: The Unnecessary Quest for Extra-Terrestrial Life in Space. Bryan Kelly. CreateSpace Independent Publishing Platform, 1st edition. Scotts Valley, CA. 2017.

Ancient Aliens on the Moon. Mike Bara. Adventures Unlimited Press. Kempton, IL. 2012.

Anunnaki, UFOs, Extraterrestrials & Afterlife Greatest Information as Revealed by De Lafayette. Maximillien de Lafayette. Times Square Press and Elite Associates. New York, NY. 2008.

The Astronomical Register, Vol. 20, Frederick William Levander, Forgotten Books, London, England, 1882

Atlantis, The Lost City. Andrew Donkin. DK Publishing, Inc. New York, NY. 2000. Library Book.

The Book of the Damned: A Collected Works of Charles Fort. Jeremy P. Tarcher/Penguin. 1919.

Chronological Catalogue of Reported Lunar Events. Barbara M. Middlehurst, Jaylee M. Burley, Patrick Moore, Barbara L. Welther. NASA. Washington, DC. 1968.

Cosmological Ice Ages. Henry Kroll. Trafford Publishing. Victoria, BC, Canada. 2009.

Dark Mission. Richard C. Hoagland and Mike Bara. Feral House. Port Townsend, WA. 2007.

Elder Gods of Antiquity. M. Don Schorn. Ozark Mountain Publishing. Huntsville, AR.. 2008.

The Encyclopedia of Moon Mysteries: Secrets, Conspiracy Theories, Anomalies, Extraterrestrials and More. Constance Victoria Briggs. Adventures Unlimited Press, Kempton, IL. 2019.

ET's Are on the Moon & Mars. C.L. Turnage. Timeless Voyager Press. Santa Barbara, CA. 2013.

Extraterrestrial Archaeology. David Hatcher Childress. Adventures Unlimited Press. Kempton, IL. 2000.

The Extraterrestrial Encyclopedia. D. Darling. Three Rivers Press. New York, NY. 2000.

The Flying Saucer Conspiracy. Major Donald E. Keyhoe. Henry Holt and CompaNY. Inc. New York, NY. 1955.

Flying Saucers on the Attack. Harold T. Wilkins. Ace Star Books. New York, NY. 1967.

For the Moon is Hollow and Aliens Rule the Sky. Rob Shelsky. GKRS Publications. San Bernardino, CA. 2014.

From Outer Space. Howard Menger. New Saucerian Books. Point Pleasant, WV. 1959.

Invader Moon: Who Brought the Moon and Why. Rob Shelsky. Permuted Press LLC. New York, NY. 2016.

Is Someone on the Moon? Ken Hudnall. Omega Press, El Paso, TX. 2018.

Leap of Faith. L. Gordon Cooper. HarperCollins Publishers. New York, NY. 2000.

The Lost Realms. Zecharia Sitchin. Bear & Company. Rochester, VT. 1990.

The Moon Watcher's Companion. Donna Henes. Marlowe & Company. New York, NY. 2004.

Moongate. William L. Brian. Future Science Research Publishing Co., Portland, OR, 1982.

Moonscapes. Rosemary Ellen Guiley. Prentice Hall Press. New York, NY. 1991.

Mysteries of the Unexplained. The Reader's Digest Association, Inc., Pleasantville, New York/Montreal, 1982.

New Atlas of the Moon. Serge Brunier. Firefly Books. Ontario, Canada. 2006.

New Lands, Charles Fort. Boni and Liveright, Inc., New York, NY. 1923.

Nothing in This Book Is True, But It's Exactly How Things Are. Bob

Frissell. North Atlantic Books. Berkeley, CA. 1994.

The Only Planet of Choice, Essential Briefings from Deep Space. Compiled by Phyllis V. Schlemmer & Palden Jenkins. Gateway Books. Bath. 1993.

Other Worlds Than Ours. C. Maxwell Cade, Taplinger Publishing Co., New York, NY. 1966.

Our Ancestors Came from Outer Space. Maurice Chatelain. Doubleday & Company. Inc. New York, NY. 1977.

Our Cosmic Ancestors. Maurice Chatelain. Light Technology Publishing. Flagstaff, AZ. 1988.

Our Mysterious Spaceship Moon. Don Wilson. Sphere Books Limited. London, England 1976.

Return to Earth. Buzz Aldrin. Open Road, Integrated Media. New York, NY. 1973.

Russian Scientists Published This Paper in the 70s: 'Is the Moon the Creation of Alien Intelligence?, https://tall-white-aliens.com/?p=12908

Secrets of Our Spaceship Moon. Don Wilson. Dell Publishing. New York, NY. 1979.

The Secret Space Program and Breakaway Civilization. Richard Dolan. Richard Dolan Press. 2016.

Somebody Else is on the Moon. George Leonard. First Published with McKay, New York, 1976. Republished Kalpaz Publications, Delhi, India. 2017.

The Source: Journey Through the Unexplained. Art Bell and Brad Steiger. New American Library. New York, NY. 2002.

The UFO Book: Encyclopedia of the Extraterrestrial. Jerome Clark. Visible Inc. Press. Detroit, MI. 1998.

Time, 70th Anniversary Celebration, 1923-1993. TIME Books. New York, NY. 1994.

To Rule the Night, The Discovery Voyage of Astronaut Jim Irwin. James B. Irwin with William A. Emerson, Jr. A.J. Holman Company. Division of J.B. Lippincott Company. Philadelphia & New York. 1973.

We Discovered Alien Bases on the Moon, II. Fred Steckling. G.A.F. International. Vista, CA. 1981.

Weird Astronomy: Tales of Unusual, Bizarre, and Other Hard to Explain Observations. David A.J. Seargent. Springer. New York, NY. 2010.

The Moon's Galactic History

Who Built the Moon? Christopher Knight and Alan Butler. Watkins Publishing. London, England. 2005.

Who Has Walked on the Moon, https://solarsystem.nasa.gov/news/890/who-has-walked-on-the-moon/

Worlds in Collision. Immanuel Velikovsky. Doubleday. New York, NY. 1951.

Image Sources

Continua La Polemica Por Nave Nodriza Descubierta En Un Crater De La Luna, https://hombresdenegronetwork.blogspot.com/2014/10/continua-la-polemica-por-nave-nodriza.html

The Gate of the Sun Calendar from Ancient Tiwanacu, https://blog.world-mysteries.com/science/the-gate-of-the-sun-calendar-from-ancient-tiwanacu/

How NASA is Taking Steps Towards a Lunar ColoNY. https://leotferguson.medium.com/how-nasa-is-taking-steps-towards-a-lunar-colony-7fa68d552b3e

NASA.gov, https://www.nasa.gov/centers/ames/news/releases/2004/moon/moon_images.html

Here Are NASA's Unreleased Apollo Mission Images They Don't Want You To See, https://humansarefree.com/2020/10/nasa-unreleased-apollo-mission-images-ufos.html

National UFO Center, Filer's Research Institute, Ret. Major George Filer, https://nationalufocenter.com/artman/publish/article_359.php

Say Hello Spaceman, A Cultural History, https://sayhellospaceman.blogspot.com/2015/12/way-way-out-1966-gallery-2.html

Satellite Dish Image, https://duckduckgo.com/?t=ffab&q=satallite+dish+on+the+Moon&atb=v278-1&iax=images&ia=images&iai=https%3A%2F%2Fi.ytimg.com%2Fvi%2FeFmusoyYcy8%2Fhqdefault.jpg

Internet Sources

10 Theoretical Megastructures, From Big to Massive, https://gizmodo.com/10-theoretical-megastructures-from-big-to-massive-1776749017

10 Things We Know About Elon Musk's Future Colony On Mars, https://listverse.com/2018/04/07/10-things-we-know-about-elon-musks-future-colony-on-mars/

1969: Astronaut Buzz Aldrin Recounts Apollo 11 UFO Encounter,

From *A Trip to the Moon*, 1902.

https://www.thinkaboutitdocs.com/1969-astronaut-buzz-aldrin-recounts-apollo-11-ufo-encounter/

52nd Lunar and Planetary Science Conference 2021 (LPI Contrib. No. 2548), The Lunar Underground: The Lava-Tube Investigation (LA-TUNA) Mission Concept (2021), https://www.hou.usra.edu/meetings/lpsc2021/pdf/1574.pdf

A Brighter Moon, http://www.varchive.org/itb/brigmoon.htm

A Cultural Roundup of the Moon, http://magazine.gwu.edu/a-cultural-roundup-of-the-moon

A facelift for the Moon every 81,000 years, http://m.phys.org/news/2016-10-facelift-moon-years.html

A Master Mathematician and His Bizarre Efforts to Contact Martians, https://mysteriousuniverse.org/2021/11/a-master-mathematician-and-his-bizarre-efforts-to-contact-martians

A Mercury Colony?, https://www.einstein-schrodinger.com/mercury_colony.html

A prominent lunar impact crater found in the southeastern region of the Moon, https://www.findagrave.com/memorial/65131128/frank-august-halstead

A Short Biography of James Clarke Greenacre, By Robert O'Connell

and Anthony Cook, http://www.the1963aristarchusevents.com/A_
Short_Biography_of_James_C__Greenacre_2013-07-28.pdf
A Time Before the Moon–How Did the Moon Get Here?, https://
www.youtube.com/watch?v=9z-DYIHDEsE
Above Top Secret, https://www.abovetopsecret.com/forum/
thread447131/pg1
Above Top Secret: NASA #2 Moon, http://www.abovetopsecret.com/
forum/thread418333/pg2
Ancient Code: Buzz Aldrin, We Were Ordered Away from the Moon,
https://www.ancient-code.com/buzz-aldrin-we-were-ordered-away-
from-the-moon/
Ancient Code, https://www.ancient-code.com/official-apollo-
mission-transcripts-reveal-fascinating-details-about-aliens/
*Ancient Origins: The origins of human beings according to ancient
Sumerian texts*, http://www.ancient-origins.net/human-origins-
folklore/origins-human-beings-according-ancient-sumerian-
texts-0065
*Alex Collier, ET Contactee Says Moon Has Vegetation and an
Atmosphere*, http://www.educatinghumanity.com/2011/11/moon-
extraterrestrial-contactee-say.html
Alex Collier Moon and Mars Lecture 1996, http://www.alexcollier.
org/3
Alien City on Moon Seen in NASA Photo, UFO Researchers Say,
http://www.inquisitr.com/2484287/nasa-alien-city-on-moon-ufo/
*Alien Conspiracy, A Different Point of View, Exposes Endymion
Lunar Operations Command Post*, http://alienconspiracy.club/
endymion-lunar-operations-command-post/
Alien Moon Bases on the Dark Side of the Moon, https://
businessecon.blogspot.com/2014/01/alien-moon-bases-on-dark-
side-of-moon.html
*Alien Presence on the Moon—Onboard Tapes and Accounts from
Those Involved*, https://anomalien.com/alien-presence-on-the-moon-
onboard-tapes-and-accounts-from-those-involved/
Aliens & the Old West, https://ancientalienpedia.com/home/part-
three/3-1-aliens-the-old-west-7-28-11/
*Aliens and the Brookings Institution: Space-Life Report Could be
Shock*, https://anomalien.com/aliens-and-the-brookings-institution-
space-life-report-could-be-shock

Aliens, Everything You Want to Know, Aliens on the Moon, http://
www.aliens-everything-you-want-to-know.com/AliensOnTheMoon.
html

Aliens Have Been on the Moon Since Ancient Times, http://ufodigest.
com/news/0708/ancient-moon2.html

Aliens on the Moon, http://sillykhan.com/tag/man-on-the-moon/

Aliens on the Moon, http://www.aliens-everything-you-want-to-
know.com/AliensOnTheMoon.html

Aliens on the Moon, http://www.nbcnews.com/science/space/aliens-
moon-tv-show-adds-weird-ufo-twists-apollo-tales-n159806

The Alternative Conquest of the Moon, https://www.
bibliotecapleyades.net/luna/esp_luna_13.htm

The Alternative Conquest of the Moon, https://www.
eyeofthepsychic.com/moon/

*The American Heritage® Dictionary of the English Language,
5th Edition,* Megastructure, https://www.wordnik.com/words/
megastructure

Anaxagoras, http://www.iep.utm.edu/anaxagor/

Ancient History Encyclopedia, Ovid, https://www.ancient.eu/ovid

Ancient, Shattered Lunar Domes, http://www.bibliotecapleyades.net/
luna/esp_luna_79.htm

*Angelismarriti.it, An Alien Spaceship on the Moon:
Interview with William Rutledge, Member of the Apollo 20
Crew,* by Luca Scantamburlo, http://www.angelismarriti.it/
ANGELISMARRITI-ENG/REPORTS_ARTICLES/Apollo20-
InterviewWithWilliamRutledge.htm

*Another Interesting Leak, A Second NASA Scientist Tells Us That
'Somebody Else' Is On the Moon,* http://www.collective-evolution.
com/2016/01/02/another-interesting-leak-a-second-nasa-scientist-
tells-us-that-somebody-else-is-on-the-moon/

Apollo 10, https://history.nasa.gov/SP-4029/Apollo_10a_Summary.htm

Apollo 10 Mission Objective: http://www.nasa.gov/mission_pages/
apollo/missions/apollo10.html#.VLbhTihBrgE

Apollo 11 Moon Landing—50 Years Later and Still Hiding the Truth?
https://endtimesand2019.wordpress.com/2019/07/19/apollo-11-
moon-landing-50-years-later-and-still-hiding-the-truth/

Apollo 13, Houston We've Got a Problem, https://er.jsc.nasa.gov/seh/
pg15.htm

Apollo 14 astronaut claims peace-loving aliens prevented 'nuclear war' on Earth, http://www.foxnews.com/science/2015/08/14/apollo-14-astronaut-claims-peace-loving-aliens-prevented-nuclear-war-on-earth.html

Apollo 20, Journey into Darkness, https://theunredacted.com/apollo-20-journey-into-darkness/

Apollo Moon Conversations and Pictures Show NASA Cover-Up, http://www.bibliotecapleyades.net/luna/esp_luna_4a.htm#Another%20strange%20Apollo%2016%20ground-to-air%

Apollo Missions, http://news.discovery.com/space/the-lost-apollo-missions-110902.htm

Apollo Moon Conversations and Pictures Show NASA Cover-up, http://www.ufos-aliens.co.uk/cosmicphotos.html

The Archimedes Platform, http://www.astrosurf.com/lunascan/luna1.html

The Arcturians, https://www.bibliotecapleyades.net/vida_alien/alien_races01a.htm

The Arcturians—The Most Evolved Alien Specie in Our Galaxy and Earth's Wardens, https://anomalien.com/the-arcturians-the-most-evolved-alien-specie-in-our-galaxy-and-earths-wardens/

The Aristarchus Anomaly: A Beacon on the Moon, http://mysteriousuniverse.org/2013/11/the-aristarchus-anomaly-a-beacon-on-the-moon/

Astonishing Intelligent Artifacts Found on Mysterious Far Side of the Moon, https://www.bibliotecapleyades.net/luna/esp_luna_35.htm

Astounding Moon Footage, http://www.bibliotecapleyades.net/luna/esp_luna_5.htm

Astronaut Gordon Cooper's letter to the UN regarding UFO, https://setiathome.berkeley.edu/forum_thread.php?id=61318

Astronaut John Young, One of Only 12 to Walk on Moon, Dies at 87, http://fortune.com/2018/01/06/astronaut-john-young-dead/

Astronaut "UFO" Sightings, http://debunker.com/texts/astronaut_ufo.html

Astronaut Walter Schirra Dies at 84-Had Reported UFO, https://www.ufocasebook.com/schirra.html

Astronauts Who Report UFOs Are Heroes, https://www.ufodigest.com/article/astronauts-who-report-ufos-are-heroes/

Astronauts Who Told the World We Are Being Visited & What They

Said, http://exonews.org/astronauts-who-told-the-world-we-are-being-visited-what-they-said/

Astronauts Who Report UFOs Are Heroes, UFO Digest, https://www.ufodigest.com/article/astronauts-who-report-ufos-are-heroes/

Astronomers Admit to Seeing Triangle UFO's Circling Moon, https://www.top10ufo.com/astronomers-admit-to-seeing-triangle-ufos-circling-moon/

Astronomers Admit to Triangles Transversing the Moon, http://worldufophotosandnews.org/?p=9905

Atlantis Online, https://atlantisonline.smfforfree2.com/index.php/topic,3238.30.html

Before the Flood, there was no Moon, http://www.pravdareport.com/news/russia/10-10-2002/13385-0/

Before the Moon Existed, https://verumetinventa.wordpress.com/2015/03/08/articles-before-the-moon-existed-part-5-by-raymond-towers/

Before the Moon, Historical Observations, https://verumetinventa2.wordpress.com/2017/06/16/before-the-moon-02-historical-observations-by-raymond-towers/

Ben Rich CEO Lockheed Skunk Works, Quotes, https://www.goodreads.com/author/quotes/14718802.Ben_Rich_CEO_Lockheed_Skunk_Works

Ben Rich, Lockheed Martin and UFOs, https://www.gaia.com/article/ben-rich-lockheed-martin-and-ufos

Bep Kororoti and the Ancient Astronauts, https://www.bibliotecapleyades.net/vida_alien/alien_AAtheory06.htm

Black Prince Alien Space Probe Orbits Earth Watching Humans, http://ufodigest.com/article/black-prince-alien-space-probe-orbits-earth-watching-humans

Borman Lowell Anders, Apollo 8, http://nssdc.gsfc.nasa.gov/planetary/lunar/apollo8info.html

Breakaway-civilization-behind-mysterious-secret-space-program, http://ufodigest.com/article

The Brookings Report and the Implications of a Discovery of Extraterrestrial Life, https://hypnagogia.wordpress.com/2016/02/01/the-brookings-report-and-the-implications-of-a-discovery-of-extraterrestrial-life/

Buzz Aldrin Confirms UFO Sighting in Syfy's 'Aliens on the Moon,' https://www.youtube.com/watch?v=ZNkmhY_ju8o

The Cameras That Recorded The Moon Landing, https://digitalrev.com/2016/07/21/the-cameras-that-recorded-the-moon-landing

Charles Duke Quotes, https://www.azquotes.com/author/37483-Charles_Duke

China Releases Photos of Structures on the Moon, By Chang'e-2 Orbiter, http://www.educatinghumanity.com/2012/02/alien-moon-base-china-releases-photos.html

The Chinese on the Moon, http://www.bibliotecapleyades.net/luna/esp_luna_79.htm

Cities Found on the Moon!, http://www.bibliotecapleyades.net/luna/luna_moonanomalies01.htm

Cities Found on the Moon!, http://www.pravdareport.com/news/russia/05-10-2002/13626-0/

Clementine Mission, https://www.nasa.gov/mission_pages/LCROSS/searchforwater/clementine.html

Colonization of Titan, https://infogalactic.com/info/Colonization_of_Titan

Colonization of Titan, https://spacecolonization.fandom.com/wiki/Colonization_of_Titan

Concerning the Face, Which Appears in the Orb of the Moon, http://penelope.uchicago.edu/Thayer/e/roman/texts/plutarch/moralia/the_face_in_the_moon*/a.html

Core Concept: Lava tubes may be havens for ancient alien life and future human explorers, https://www.pnas.org/content/117/30/17461

Coyolxauhqui, https://www.windows2universe.org/mythology/coyolxauhqui_moon.html

Craters on the Edge, http://www.esa.int/spaceinimages/Images/2007/02/Craters_on_the_edge

Craters, http://www.bibliotecapleyades.net/ciencia/ciencia_nemesis08.htm

The Daily Galaxy, Earth's Secrets to be kept in a Lunar "Noah's Ark," http://www.dailygalaxy.com/my_weblog/2008/03/mankinds-secret.html

Dark Side of the Moon, http://houseoftantra.org/the-moon/

DavidIcke.com, http://forum.davidicke.com/showthread.php?t=108665

Did Aliens Help Apollo 13 Return to Earth?, http://www.

thecryptocrew.com/2012/01/did-aliens-help-apollo-13-return-to.html

Did Angels Interbreed With Women to Produce Giants?, https://www.
ucg.org/bible-study-tools/booklets/angels-gods-messengers-and-
spirit-army/did-angels-interbreed-with-women-to-produce-giants

Did Apollo 17 find a Stargate on the Moon?, http://exonews.org/
apollo-17-find-stargate-moon-exonews-tv-s01e13/

Did Apollo 17 find a Stargate on the Moon, http://www.
bibliotecapleyades.net/stargate/stargate24.htm

Did Atlantis Really Exist?, https://earthsky.org/earth/lost-continent-
zealandia-drilling-expedition-2017

Dr. Zaius, https://villains.fandom.com/wiki/Dr._Zaius

The Earth without the Moon, http://www.varchive.org/itb/sansmoon.
htm

Edgar Mitchell, https://www.azquotes.com/quote/606633

Encounter with Jesus on the Moon Left Astronaut Changed, http://
blog.godreports.com/2011/03/encounter-with-jesus-on-the-moon-
left-astronaut-changed/

The Enigmas on the Moon, http://www.thelivingmoon.
com/43ancients/02files/Moon_Images_Book02.html

Exemplore, *Ancient Cities Found on the Moon*, Dylan Clearfield
(2020), https://exemplore.com/ufos-aliens/Alien-Threat-from-the-
Moon.

False Crater Floors, Plato Crater, http://www.thelivingmoon.
com/43ancients/41Group_Lunar_FYEO/02files/FYEO_Lunar_04a.
html

*The Fascinating Story Behind the Most Famous Picture of Earth
Ever Taken*, https://www.businessinsider.com/apollo-8-earthrise-
image-2013-12

Fastwalkers and Slowwalkers (Real Proof of UFOs?), http://www.
abovetopsecret.com/forum/thread183369/pg1

The First Photograph of the Moon, http://time.com/3805947/the-
first-photograph-of-the-moon/

Footprints on the Moon, http://www.think-aboutit.com/Moon/
FootprintsontheMoon.htm

Forbidden History, David Icke, http://www.unariunwisdom.com/
forbidden-history/

*Former Apollo astronaut claimed ancient alien astronauts created
humans evidence in Sumerian texts*, https://anciently.net/former-

apollo-astronaut-claimed-ancient-alien-astronauts-created-humans-evidence-in-sumerian-texts-10107/

Former Israeli space security chief says extraterrestrials exist, and Trump knows about it, https://www.nbcnews.com/news/weird-news/former-israeli-space-security-chief-says-extraterrestrials-exist-trump-knows-n1250333

The Galactic Federation, http://thegreaterpicture.com/galactic-federation.php

Gemini 7, https://nssdc.gsfc.nasa.gov/nmc/spacecraft/display.action?id=1965-100A

Giant Cylindrical Objects Photographed on Apollo 9 Mission, http://www.theblackvault.com/casefiles/giant-cylindrical-objects-photographed-on-apollo-9-mission/

Giant Hominoids, http://www.zetatalk.com/index/blog0926.htm

Giordano Bruno, http://www.famousphilosophers.org/giordano-bruno/

The glass domes of the moon, https://www.unexplained-mysteries.com/column.php?id=212614

Government UFO Cover Ups, https://futurism.media/government-ufo-cover-ups

The Great UFO Cover-up, http://www.youtube.com/watch?v=AbgHyrmgRZM&feature=player_embedded#

Greys, http://arcturi.com/GreyArchives/theyarecoming.html

Has Anyone Heard of the Hollow Moon Theory?, http://www.abovetopsecret.com/forum/thread212580/pg4&mem

Has China Made Contact With Aliens on the Moon?, http://ufodigest.com/article/super-tlp-captured-chinese-moon-photos-rainbow-ufo-buzzed-chinese-lunar-lander-chang-e-3

Here Are NASA's Unreleased Apollo Mission Images They Don't Want You To See, https://humansbefree.com/2020/10/nasa-unreleased-apollo-mission-images-ufos.html

Historical Events in 1968, http://www.historyorb.com/events/date/1968

Historical Review of Lunar Anomalies, http://www.cavinessreport.com/ad3_3.html

History: Atlantis, http://www.history.com/topics/atlantis

History Channel, Joseph Walker NASA Astronaut UFO Sighting, https://www.bing.com/videos/search/Joseph+Walker+NASA+pilot

History Disclosure, http://www.historydisclosure.com/there-was-a-time-when-the-moon-did-not-exist/

The History of Earth, http://thegreaterpicture.com/earth.html

How Apollo 11s moonquakes changed our understanding of earthquakes, https://temblor.net/earthquake-insights/how-apollo-11s-moonquakes-changed-our-understanding-of-earthquakes-9213/

How is the Moon Classified?, https://sciencing.com/moon-classified-22799.html

How much Trash is on the Moon, https://www.livescience.com/61911-trash-on-moon.html

How to build a moon base, https://theconversation.com/how-to-build-a-moon-base-120259

Hugh Percival Wilkins, 1896-1960: an Appreciation, http://articles.adsabs.harvard.edu/cgi-bin/nph-iarticle_query

Human Evolution After Colonizing the Moon, https://sservi.nasa.gov/?question=human-evolution-after-colonizing-the-moon

Humanoid, https://www.merriam-webster.com/dictionary/humanoid

Humans are Free: Who 'Parked' the Moon in Perfect Circular Orbit Around Earth?, http://humansarefree.com/2013/12/who-parked-moon-in-perfect-circular.html

If We Discover Aliens, What's Our Protocol for Making Contact?, https://www.livescience.com/19360-humans-discover-aliens.html

Immense triangular-shaped lighted object discovered on surface of Moon, https://www.altereddimensions.net/2014/huge-triangular-shaped-lighted-object-discovered-on-surface-of-moon

The Incredible Shrinking Moon, http://www.dailymail.co.uk/sciencetech/article-1304627/Cracks-surface-Moon-reveal-closest-neighbour-actually-shrinking.html

Inexplicable Moon Anomalies Reveal Deception about Outer Space, http://www.wakingtimes.com/2015/10/19/inexplicable-moon-anomalies-reveal-deception-about-outer-space/

Ingo Swann Gets Feedback Regarding Naked Men on the Moon, http://www.bibliotecapleyades.net/vision_remota/esp_visionremota_penetration

Ingo Swann's Remote Viewing of Jupiter, http://ersby.blogspot.com/2016/07/ingo-swanns-remote-viewing-of-jupiter.html

Interview with Alex Collier, http://www.bibliotecapleyades.net/andromeda/esp_andromedacom_8.htm

Iopscience, http://iopscience.iop.org/article/10.1086/121039/pdf

Is Earth's Moon Really a Natural Satellite? (Astronomical Observations & Secret Data), https://www.youtube.com/watch?v=zo9DScR1B3A&fbclid=IwAR3Ih3Z-SVpXxYwAsW6SX1-j_eEdQ4Qz58_zMFFjF9nYhQvHcX6_BS3Z5uo

Is NASA Gradually Revealing Evidence of Extraterrestrial Life on the Moon? https://www.gaia.com/article/nasa-disclosing-alien-life-on-moon-and-mars?render=details-v4&utm_source=google+paid&utm_medium=cpc&utm_term=&utm_campaign=1-USA-DYNAMIC-SEARCH&utm_content=UFOs+&+Aliens&ch=br&gclid=Cj0KCQiA3Y-ABhCnARIsAKYDH7sE85xrfTlA30s7sHNNWoO5m7lhf0JU2pAWuzdj3FcoiChofEnccZ8aAlXjEALw_wcB

Is the Moon a Planet?, http://www.universetoday.com/85749/is-the-moon-a-planet/

Is the Moon Real? New Analysis Shows it Might be a HOLOGRAM, https://www.disclose.tv/is-the-moon-real-new-analysis-shows-it-might-be-a-hologram-312854

Is the Moon the Creation of Intelligence?, http://www.bibliotecapleyades.net/luna/esp_luna_6.htm

Is Titan habitable?, https://guruquestion.com/is-titan-habitable

Isaac Asimov, http://www.asimovonline.com/asimov_home_page.html

The Isaac Asimov Interview, https://www.motherearthnews.com/sustainable-living/nature-and-environment/science-technology-isaac-asimov-zmaz80sozraw/

James Irwin Quotes, https://quotepark.com/authors/james-irwin/quotes-about-god/

James Lovell and Frank Borman, https://archive.ronrecord.com/astronauts/lovell-borman.html

John F. Kennedy Moon Speech - Rice Stadium, https://er.jsc.nasa.gov/seh/ricetalk.htm

John F. Kennedy Moon Speech Transcript: "We Choose to Go to the Moon," https://www.rev.com/blog/transcripts/john-f-kennedy-jfk-moon-speech-transcript-we-choose-to-go-to-the-moon

John Lear's Files, The Case for the Civilization on the Moon, http://www.thelivingmoon.com/47john_lear/02files/Case_for_Civilization_on_the_Moon.html

The Kardashev Scale: Classifying alien civilizations, https://www.

space.com/kardashev-scale

The Kardashev Scale—Type I, II, III, IV & V Civilization, https://futurism.com/the-kardashev-scale-type-i-ii-iii-iv-v-civilization

Kentwired.com, Opinion: Astronauts talk about UFOs and aliens, http://www.kentwired.com/opinion/article_8c00d9ec-e3e5-5b69-b03c-e5afd34d5cfe.html

The Lemurians and Atlantis Civilizations, https://www.bibliotecapleyades.net/atlantida_mu/esp_lemuria_8.htm

The Living Moon, https://www.thelivingmoon.com/43ancients/02files/Moon_Images.html

Living Underground on the Moon: How Lava Tubes Could Aid Lunar Colonization, https://www.space.com/moon-colonists-lunar-lava-tubes.html

The Lost Continent of Atlantis, http://www.bibliotecapleyades.net/atlantida_mu/esp_atlantida_5a.htm

Luna, http://www.bibliotecapleyades.net/luna/esp_luna_4a.htm

Luna 9: 1ˢᵗ Soft Landing on the Moon, https://www.space.com/35116-luna-9.html

Lunar and Planetary Institute, https://www.lpi.usra.edu/lunar/missions/apollo/

Lunar Mysteries, http://www.bibliotecapleyades.net/luna/esp_luna_34.htm

Lunar Orbiter 2, https://nssdc.gsfc.nasa.gov/nmc/spacecraftDisplay.do?id=1966-100A

Lunatics, http://www.halexandria.org/dward199.htm

Man Moon Mystery Solved, https://www.dailymail.co.uk/sciencetech/article-2725678/Man-moon-mystery-solved-Nasa-claims-strange-figure-just-scratch-negative-film.html

Mare Crisium, https://www.nasa.gov/mission_pages/LRO/multimedia/lroimages/lola-20100702-crisium.html

Michael Salla: Moon is Artificial & Arrived With Refugees From Destroyed Planet in Asteroid Belt, https://www.youtube.com/watch?v=HN9xAMna_AI

Mid-century Contactees who claimed to meet 'Nordic aliens' in the Mojave Desert, https://www.ancient-code.com/mid-century-contactees-who-claimed-to-meet-nordic-aliens-in-the-mojave-desert/

Military Secrets and Advances, Including Space Related Topics, http://www.thelivingmoon.com/45jack_files/index.html#Russian_

Connection

Mona Lisa, An Alien Spaceship on the Moon, Viewzone.com, https://www.youtube.com/watch?v=vkdwrfJUfm8

The Moon: A Natural Satellite Cannot Be a Hollow Object, http://subrealism.blogspot.com/2017/09/the-moon-natural-satellite-cannot-be.html

The Moon: An Unexplained Phenomenon, https://redice.tv/news/the-moon-an-unexplained-phenomenon

Moon Anomalies, http://truereality.org/moon-anomalies/

The Moon Created by Aliens?, http://www.abovetopsecret.com/forum/thread174073/pg1

The Moon Created by Aliens, Was it Built for Us?, http://www.abovetopsecret.com/forum/thread174073/pg1

The Moon, Creation of Intelligence?, http://www.bibliotecapleyades.net/luna/esp_luna_6.htm

Moon, Early Studies, http://www.britannica.com/topic/Moon-exploration: https://www.cnbc.com/2017/03/09/china-developing-manned-space-mission-to-the-moon.html

Moon Goddess, https://www.goddess-guide.com/moon-goddess.html

Moon Gods and Moon Goddesses, https://www.thoughtco.com/moon-gods-and-moon-goddesses-120395

The Moon, http://thegreaterpicture.com/earth.html

The Moon is a Foreign Nation, http://www.bibliotecapleyades.net/vida_alien/esp_vida_alien_37.htm

The Moon is a Spacecraft, http://wariscrime.com/new/the-moon-is-a-spacecraft/

The Moon is Owned by the Anunnaki and is Nibiru, http://bogusgw.blogspot.com/2010/01/moon-is-owned-by-anunnaki-and-is-nibiru.html

Moon Landing Hoax, https://news.nationalgeographic.com/news/2009/07/photogalleries/apollo-moon-landing-hoax-pictures/

Moon: Myths and Legends, http://www.mythencyclopedia.com/Mi-Ni/Moon.html

Moon Photos NASA Doesn't Want You to See, https://www.youtube.com/watch?v=5SW2UxwkL7E

The Most Evolved ET Species In Our Galaxy And Earth's Wardens Is The Arcturians, https://timefordisclosure.com/the-most-evolved-et-species-in-our-galaxy-and-earths-wardens-is-the-arcturians/

Museum of the Moon, https://my-moon.org/research/

Mysterious "Monuments" on the Moon, Argosy Magazine, http://www.astrosurf.com/lunascan/argosy_cuspids.htm

Mysterious "Monuments" on the Moon, Argosy Magazine, https://www.scribd.com/document/335958809/MYSTERIOUS-MONUMENTS-ON-THE-MOON-by-Ivan-T-Sanderson

The Mysterious Nazca Lines, https://discover.hubpages.com/education/The-Mysterious-Nazca-Lines

Mysterious Nazca Lines Discovered on the Moon, http://ufosightingshotspot.blogspot.com/2016/11/mysterious-nazca-lines-discovered-on.html

Mystery of Ancient Structures & So-called Footprints of the Gods on Moon, https://www.howandwhys.com/mystery-of-ancient-structures-so-called-footprints-of-the-gods-on-moon/

The Mystery of Aristarchus Crater, A Fusion Reactor?, http://beforeitsnews.com/space/2012/07/the-alien-fusion-reactor-on-the-moon-2421989.html

The Mystery of Phobos: A 'Hollow' Satellite Orbiting Mars?, https://www.ancient-code.com/the-mystery-of-phobos-a-hollow-satellite-orbiting-mars/

The Myth of NASA Insider Otto Binder, http://www.jamesoberg.com/binder.otto.pdf

Mystery of the Great Sphinx found in the Tycho Crater on the Moon, http://ufosightingshotspot.blogspot.com/2016/10/mystery-of-great-sphinx-found-in-tycho.html

NASA, Artemis Moon Rover's Wheels are Ready to Roll (2022), https://www.nasa.gov/feature/ames/artemis-moon-rover-s-wheels-are-ready-to-roll

NASA Astronauts Disclose UFO Encounters, Open Minds, http://www.openminds.tv/nasa-astronauts-disclose-ufo-encounters/1476

NASA, Gemini 9A, https://nssdc.gsfc.nasa.gov/nmc/spacecraftDisplay.do?id=1966-047A

NASA Image and Video Library, https://images.nasa.gov/

NASA Image Shows Hostile Alien UFO Escorting Apollo 17 Mission Crew Off the Moon After Warning Them Not To Return, Conspiracy Theorists Claim, https://www.inquisitr.com/3159631/nasa-image-shows-hostile-alien-ufo-escorting-apollo-17-mission-crew-off-the-moon-after-warning-nasa-not-to-return-conspiracy-theorists-claim-video/

NASA Johnson Space Center Oral History Project, Edited Oral History Transcript, James A. McDivitt, https://www.jsc.nasa.gov/history/oral_histories/McDivittJA/mcdivittja.htm

The NASA Lies and Tesla Cover-Ups, http://www.bibliotecapleyades.net/luna/esp_luna_29.htm

NASA, Lunar Orbiter 2, https://nssdc.gsfc.nasa.gov/nmc/spacecraftDisplay.do?id=1966-100A

NASA, Lunar Reconnaissance Orbiter: Gassendi's Fractures, https://www.nasa.gov/mission_pages/LRO/multimedia/lroimages/lroc-20101104-fractures.html

NASA, Lunar Rocks and Soils from Apollo Missions, https://curator.jsc.NASA.gov/lunar

NASA: Lyndon B. Johnson Space Center, Frank Borman, NASA Astronaut (Former), https://www.jsc.nasa.gov/Bios/htmlbios/borman-f.html

NASA Might Put a Huge Telescope on the Far Side of the Moon (2021), https://www.wired.com/story/nasa-might-put-a-huge-telescope-on-the-far-side-of-the-moon/

NASA Photographed Gigantic Cylindrical UFO In Front Of Moon, http://www.theblackvault.com/casefiles/category/space-anomalies/moon-anomalies/

NASA Photographed Gigantic Cylindrical UFO In Front Of Moon, http://www.ufo-blogger.com/2013/10/NASA-photographed-cylindrical-UFO-in-front-of-moon.html

NASA Releases Recording of Strange 'space music' heard by Apollo 10 astronauts, http://www.telegraph.co.uk/news/science/space/12169511/Nasa-releases-recording-of-strange-space-music-heard-by-Apollo-10-astronauts.html

NASA Space Science Data Coordinated Archive: Zond 3, NSSDCA/COSPAR ID: 1965-056A, https://nssdc.gsfc.nasa.gov/nmc/spacecraftDisplay.do?id=1965-056A

NASA, The Future, https://www.nasa.gov/specials/60counting/future.html

NASA, UFO No Longer Unidentified, https://www.nasa.gov/vision/space/travelinginspace/no_ufo.html

National Aeronautics and Space Administration: HumanSpaceFlight, https://spaceflight.nasa.gov/gallery/images/apollo/apollo7/ndxpage1.html

Nazca Lines, http://answers.askkids.com/Old_Stuff/what_are_the_
nazca_lines

Nazca Lines, https://earth-chronicles.com/histori/the-patterns-of-the-
nazca-lines-repeat-patterns-on-the-surface-of-the-moon.html

*NBC News, Former Israeli space security chief says extraterrestrials
exist, and Trump knows about it*, https://www.nbcnews.com/news/
weird-news/former-israeli-space-security-chief-says-extraterrestrials-
exist-trump-knows-n1250333

*NBC News, UFOs and Beyond, Apollo 14 Astronaut Edgar Mitchell
is Looking Up*, http://www.nbcnews.com/science/space/ufos-beyond-
apollo-14-astronaut-ed-mitchell-looking-n165321

*Neil Armstrong Moon Landing and Presence Of UFO's and Aliens
On Moon:* http://www.ufo-blogger.com/2009/07/neil-armstrong-
moon-landing-and.html

Neil Armstrong Saw Aliens on Moon Proof, https://www.youtube.
com/watch?v=ccGqU0eczQg

*Northern Ontario UFO Research & Study (NOUFORS), Official
UFO Documents*, http://www.noufors.com/official_ufo_documents.
html.

NOUFORS, http://www.noufors.com/images/nasa_spacelife.jpg

The Official Website of Zecharia Sitchin, https://www.sitchin.com

*One of the earliest pictures of the moon by Dr. J. W. Draper of New
York*, 1840, https://trending.com/tweets/2019-02-02/one-of-the-
earliest-pictures-of-the-moon-by-dr-j-w-draper-of-new-york-1840

Ovid: Fasti, Book Two, The Lupercalia, https://www.
poetryintranslation.com/PITBR/Latin/OvidFastiBkTwo.php

Penetration - The Question of Extraterrestrial and Human Telepathy,
http://www.bibliotecapleyades.net/vision_remota/esp_visionremota_
penetration.htm#Penetration:%20The%20Question%20of%20
Extraterrestrial%20and%20Human%20Telepathy

Peninsula Clarion, All About the Moon, http://peninsulaclarion.com/
outdoors/2013-03-28/all-about-the-moon

Peruvian Stargate, https://www.crystalinks.com/perustargate.html

Photos: Mysterious Objects Spotted on the Moon, http://www.
livescience.com/33883-gallery-weird-moon.html

Pierre-Simon, Marquis De Laplace, https://www.britannica.com/
biography/Pierre-Simon-marquis-de-Laplace

Planet of the Apes Quotes, https://www.rottentomatoes.

com/m/1016397_planet_of_the_apes/quotes

Planet X Under Our Nose? Has the Moon hit the Earth in the past?, http://projectavalon.net/forum4/showthread.php?74082-Planet-X-under-our-nose-Has-the-Moon-hit-the-Earth-in-the-past

Plutarch: Greek Biographer, https://www.britannica.com/biography/Plutarch

Plutarch, Stanford Encyclopedia of Philosophy, (2010, 2014) https://plato.stanford.edu/entries/plutarch

Prabook, Viktor Mikhailovich Afanasyev, https://prabook.com/web/viktor_mikhailovich.afanasyev/3715123

Preliminary Search for Ruin-Like Formations on the Moon, http://www.astrosurf.com/lunascan/arkhipov2.htm

Quotes About Space Exploration, https://www.quotemaster.org/Space+Exploration

Reptilians, https://news-intel.com/the-moon-was-brought-reptilian/

Restoration of American Heroes: Astronauts, Who Report About Extraterrestrial Encounters, http://www.bibliotecapleyades.net/luna/luna_apollomissions07.htm

The Round Robin, October 1946, No. 10 of Volume 2, https://www.joshuablubuhs.com/blog/newton-meade-layne-as-fortean

Salvation of Humans.com, An Alien Colony Banned to Humans?, http://www.salvation-of-humans.com/English/05-01_e_t_and-lies.htm#colonie_interdite

Scariest Book of all Time, http://scariestbookofalltime.blogspot.com/2012/04/cosmological-ice-ages-short-explanation.html

Scientist Claims Mars Moon Phobos Is Hollow and Artificial!, https://www.auricmedia.net/scientist-claims-mars-moon-phobos-is-hollow-and-artificial/

Scientists demonstrate how we can live on the moon, https://www.thebrighterside.news/post/scientists-demonstrate-how-we-can-live-on-the-moon

Scientists reveal design plans for future lunar base, https://sservi.nasa.gov/articles/scientists-reveal-design-plan-for-future-lunar-base/

Scientists Suggest Moon Photos May Reveal Extraterrestrial Visitation, http://www.huffingtonpost.com/alejandro-rojas/scientists-suggest-moon-p_b_1173124.html

Sea of Tranquility, https://www.universetoday.com/50525/sea-of-tranquility/

Secrets of Our Spaceship Moon, https://www.scribd.com/
doc/40405065/Secrets-of-Our-Spacecship-Moon
The Secrets of Schroteri, http://www.thelivingmoon.
com/43ancients/02files/Moon_Images_A30_Shcroteri_Crater.html
Selenites, http://www.setileague.org/reviews/selenite.htm
Sensation: Cities Found on the Moon!, http://www.pravdareport.
com/news/russia/05-10-2002/13626-0/#.VKMzUihBrgE
*Sensational Claim: US Air Force whistleblower says 'I saw
structures on dark side of moon',* https://www.express.co.uk/
news/weird/688822/SENSATIONAL-CLAIM-US-Air-Force-
whistleblower-saw-structures-on-dark-side-of-moon
The SETI League's Volunteer Coordinator for Ukraine, http://www.
setileague.org/admin/alexey.htm
The Shard, http://www.thelivingmoon.com/43ancients/02files/Moon_
Others_01.html#Shard
Sitchen Studies, The Anunnaki, https://www.sitchinstudies.com/the-
anunnaki.html
*Sixth man to walk on the moon says aliens prevented nuclear war
between US and Russia, helped create peace on Earth,* http://www.
nydailynews.com/news/national/sixth-man-walk-moon-aliens-
prevented-nuclear-war-article-1.2324216.
*Smithsonian National Air and Space Museum, The Apollo Program:
Apollo 8,* http://airandspace.si.edu/explore-and-learn/topics/apollo/
apollo-program/orbital-missions/apollo8.cfm
*Something New Under the Sun: Crystal Clark Interviews James
Horak,* https://drowninginabsurdity.org/tag/nasa/
*Space Elevators On Hold At Least Until Stronger Materials Are
Available, Experts Say,* https://www.huffpost.com/entry/space-
elevators-stronger-materials_n_3353697
Space Encounters, 1965: The Gemini 7 UFO Sighting, https://www.
thinkaboutitdocs.com/1965-the-gemini-7-ufo-sighting/
Space-Life Report Could Be Shock, http://www.noufors.com/images/
nasa_spacelife.jpg
Spots on the Moon, Time Magazine, http://content.time.com/time/
subscriber/article/0,33009,870631,00.html
Stargate, https://www.definitions.net/definition/Stargate
Strange Attraction, Howard Menger's Final Journey, http://
strangeattractor.co.uk/news/howard-mengers-final-journey/

Sumerianology, https://sumeria.fandom.com/wiki/Sumerian_King_List

Tank on the Moon, http://www.bibliotecapleyades.net/luna/esp_luna_64.htm

Theodorus of Cyrene, http://www-history.mcs.st-andrews.ac.uk/Biographies/Theodorus.html

There is an Alien Base on the Moon, https://www.youtube.com/watch?v=aAuMkMFaYrk

There was a Time When the Moon Did Not Exist, http://www.historydisclosure.com/there-was-a-time-when-the-moon-did-not-exist/

These are the 12 things most likely to destroy the world, https://www.vox.com/2015/2/19/8069533/end-of-the-world

Think About It: Alien Type Summary—Moon Eyes, October 15, 2012, https://thinkaboutit.site/aliens/moon-eyes/

Tiamat Planetary Theory, https://www.tokenrock.com/explain-tiamat-planetary-theory-144.html

Time Before the Moon, https://www.linkedin.com/pulse/time-before-moon-lord-edwin-e-hitti

To the Moon and Back, http://www.bibliotecapleyades.net/ciencia/ciencia_psycho04.htm

Towards Lunar Archaeology, Dr. Alexey V. Arkhipov, http://www.thelivingmoon.com/43ancients/02files/Mars_Images_24.html#Towards

Transient Lunar Phenomenon (TLP), http://www.the1http://adsabs.harvard.edu/full/1992JBAA..102..157B

The Triesnecker Pyramid, http://www.thelivingmoon.com/43ancients/02files/Moon_Others_02.html#Triesnecker

The Truth About Alex Collier, https://exemplore.com/ufos-aliens/Alex-Collier-August-2012-The-Man-Behind-The-Myth-Letters-From-Andromeda?li_source=LI&li_medium=m2m-rcw-exemplore

The Truth Behind Apollo 18, http://alienxfiles.com/the-truth-behind-apollo-18

The Truth Denied: UFO Bases on the Moon & Dark Satellite Footage, http://www.thetruthdenied.com/news/2014/07/04/ufo-bases-on-the-moon-dark-satellite-footage-2014/

Time Before the Moon, Lord Edwin E. Hitti, (2019), https://www.linkedin.com/pulse/time-before-moon-lord-edwin-e-hitti

Tycho and Copernicus: Lunar Ray Craters, https://apod.nasa.gov/apod/ap010809.html

Type 2 Lunar Civilization, http://www.bibliotecapleyades.net/luna/esp_luna_79.htm

U.F.O. and Reported Extraterrestrials on Moon and Mars, http://www.bibliotecapleyades.net/vida_alien/esp_vida_alien_37.htm

The UFO Briefing Document Quotations, https://www.bibliotecapleyades.net/ciencia/ufo_briefingdocument/quoastro.htm

UFO Case Book: Alien Presence on the Moon, http://www.ufocasebook.com/moon.html

UFO Cover-Ups Must End, Moonwalker Edgar Mitchell Says, http://www.bloomberg.com/news/2013-07-16/ufo-cover-ups-must-end-moonwalker-edgar-mitchell-says.html

UFO Coverup, An Andromedan Perspective of Our, Moon, Alex Collier, http://www.oocities.org/marksrealm/coverup023.html

UFO Evidence: Scientific Study of the UFO Phenomenon and the Search for Extraterrestrial Life, http://www.ufoevidence.org/cases/case396.htm

UFO Expert Calls for Probe After Declassified Russian Files Reveal Alien Encounters, http://exonews.org/tag/vladimir-azhazha/

UFO Expert Says He Has '100% Indisputable Proof' Alien Cities Exist On The Earth's Moon, https://brobible.com/culture/article/ufo-expert-proof-alien-cities-moon/

UFO Investigator, http://www.noufors.com/images/nasa_spacelife.jpg

UFO Sightings by Astronauts, http://pages.suddenlink.net/anomalousimages/images/astroufo.html#Conrad

UFO Sightings by Astronauts, http://www.syti.net/UFOSightings.html

UFO Sightings Daily, http://www.ufosightingsdaily.com/2015/01/alien-ship-seen-close-up-on-apollo-7.html

UFO Spotted During Apollo 7, Astronauts Redact Truth With Duct Tape, http://www.huffingtonpost.co.uk/2015/01/19/ufo-spotted-apollo-7-mission_n_6499124.html

Ufology: Lunar Anomalies by Richard Hoagland, http://www.tarrdaniel.com/documents/Ufology/lunar_anomaly.html

UFOs Spotted in 1966 NASA Project Gemini Photo—Aliens Monitoring?, http://www.digitaljournal.com/internet/alien-

ufos-monitor-1966-nasa-project-gemini-space-mission-flight/
article/423367#ixzz5ZbmrJyzQhttp://www.digitaljournal.com/
internet/alien-ufos-monitor-1966-nasa-project-gemini-space-mission-
flight/article/423367

*U.S. Astronauts Reveal Encounters With Apparent Extraterrestrials
and UFOs*, https://www.bibliotecapleyades.net/luna/luna_
apollomissions05.htm

Wake up, folks. The Moon is fake, https://crberryauthor.
com/2017/10/12/wake-up-folks-the-moon-is-fake

Was the Moon Set in Place 4.53 Billion Years Ago?, http://www.
ashtarcommandcrew.net/forum/topics/moon-2?commentId=2859786
%3AComment%3A3047425#ixzz4b9NuQKMs

Webster's Dictionary, https://www.merriam-webster.com/dictionary/
Pythagorean%20theorem

Welcome? Kareeta!, https://borderlandsciences.org/journal/vol/02/
n10/Welcome_Kareeta.html

Were Aliens Watching Apollo 12 Astronauts on the Moon?, http://
www.chron.com/news/nation-world/space/article/Were-aliens-
watching-Apollo-12-astronauts-on-the-6018034.php

Were US Astronauts Ordered Not to Report UFOs & Aliens?, http://
www.rense.com/general70/rep.htm

What Are These Strange Things on the Moon?, http://paranormal.
about.com/od/lunaranomalies/ig/Strange-Things-on-the-Moon/The-
Shard.htm

*What Strange and Frightening Discoveries did Our Astronauts Make
on the Moon,* http://www.paranoiamagazine.com/2017/08/strange-
frightening-discoveries-astronauts-make-moon/

What Would Happen If There Were No Moon? https://
insidescience.org/video/what-would-happen-if-there-were-no-
moon#:~:text=Without%20the%20moon%2C%20a%20day,go%20
by%20in%20a%20blink.

When the Earth was Moonless, http://www.halexandria.org/
dward200.htm

Who are the Tartar People?, https://www.worldatlas.com/articles/
who-are-the-tartar-people.html

Who Built the Hollow Moon an African Legend, https://www.
youtube.com/watch?v=XDpPPo2n0c4

Who 'Parked' the Moon in Perfect Circular Orbit Around Earth?,

http://www.bibliotecapleyades.net/luna/esp_luna_34.htm
Why NASA Never Returned to the Moon, http://conspiracy-watch.org/why-nasa-never-returned-to-moon/
Why We Should Build Cloud Cities on Venus, https://www.vice.com/en/article/539jj5/why-we-should-build-cloud-cities-on-venus
William Herschel's Belief in Extraterrestrial Life, http://ferrebeekeeper.wordpress.com/2010/08/25/sir-william-herschels-belief-in-extraterrestrial-life/

Magazine Sources
Argosy Magazine. Mysterious "Monuments" on the Moon. Volume 371, Number 2. August 1970.
Life Magazine. Our Journey to the Moon: Written by the Astronauts, Frank, Borman, Jim Lovell, Bill Anders. January 17, 1969.

Video Sources
10 Space Photos That Will Give You Nightmares, https://www.youtube.com/watch?v=pb2MkzCgQ-4
Apollo Space Program, https://www.youtube.com/watch?v=rK1M87sd7eg
Asteroid Mining, https://www.youtube.com/watch?v=3-3DjxhGaUg
Did Apollo 17 find a Stargate on the Moon? S01E13, https://www.youtube.com/watch?v=yd-c2xw1Swc
Dr. KC Houseman Log 301: Is the Moon a Megastructure? https://www.youtube.com/watch?v=49mVLjfSp50
The Earth Without the Moon, https://www.varchive.org/itb/sansmoon.htm#f_9
Is Earth's Moon Really a Natural Satellite? (Astronomical Observations & Secret Data), https://www.youtube.com/watch?v=zo9DScR1B3A&fbclid=IwAR3Ih3Z-SVpXxYwAsW6SX1-j_eEdQ4Qz58_zMFFjF9nYhQvHcX6_BS3Z5uo
Earth without Moon, https://www.youtube.com/watch?v=5JgzlQP9NY8
Earthfiles, Dec 9, 2020—Report on Prof. Eshed, Ph.D. & Underground Mars Base and Chat Q&A, Linda Moulton Howe, https://www.youtube.com/watch?v=UUrKesBJyFg
The Kardashev Scale With Michio Kaku: Can We Become a Type 1 Civilization? https://www.youtube.com/watch?v=PxwPfPWrOCA

The Moon's Galactic History

Moon Crater: Cities, https://www.youtube.com/watch?v=4cv3SjVK-n0

The Moon Pictures—Amazing Anomalies On The Moon—Amazing Moon Photos - Broadcast Team Alpha. Written by Tom Schaefer. Narrators Nori Love and Aage Nost of Broadcast Team Alpha. 2019

"Most People Don't Even Realize What's Coming," Elon Musk Shocking Speech (2021), https://www.youtube.com/watch?v=d2aujK2HZkI

Naked Science—Moon Mysteries, (2016), https://www.youtube.com/watch?v=0tqgWuSIZUg

Orion's Belt, Pyramids of Egypt Alignment with Orion's Belt, https://blackhistory247.wordpress.com/2020/03/28/pyramids-of-egypt-alignment-with-orions-belt

Our Moon's Forbidden History, Alex Collier, https://www.youtube.com/watch?v=E83y8URqTpk

Science Channel: YouTube, Outer Space Music Pt 1 of 2 / NASA's Unexplained Files, https://www.youtube.com/watch?v=bjLZBrQ-Oq4

Star Trek: The Next Generation. Season 6 Ep. 20. *The Chase Monologue*. AJLaw98765. https://www.youtube.com/watch?v=k-j69iVReEU.

Strange Flashing Lights Sighted on the Moon, https://www.youtube.com/watch?v=Lju6TeQlJR8.

What Alien Structures Look Like On The Moon, Bruce Sees All, https://www.youtube.com/watch?v=plpfnaipDRY

What really happened on the moon—Linda Moulton Howe. The Awakening Archive, https://www.youtube.com/watch?v=ihUTj-Do72c

Who Built the Moon?, http://www.whobuiltthemoon.com/the-moon-is-not-a-natural-planet.html

HAUNEBU: THE SECRET FILES
The Greatest UFO Secret of All Time
By David Hatcher Childress

Childress brings us the incredible tale of the German flying disk known as the Haunebu. Although rumors of German flying disks have been around since the late years of WWII it was not until 1989 when a German researcher named Ralf Ettl living in London received an anonymous packet of photographs and documents concerning the planning and development of at least three types of unusual craft. Chapters include: A Saucer Full of Secrets; WWII as an Oil War; A Saucer Called Vril; Secret Cities of the Black Sun; The Strange World of Miguel Serrano; Set the Controls for the Heart of the Sun; Dark Side of the Moon: more. Includes a 16-page color section. Over 120 photographs and diagrams.

352 Pages. 6x9 Paperback. Illustrated. $22.00 Code: HBU

HIDDEN AGENDA
NASA and the Secret Space Program
By Mike Bara

Bara delves into secret bases on the Moon, and exploring the many other rumors surrounding the military's secret projects in space. On June 8, 1959, a group at the ABMA produced for the US Department of the Army a report entitled Project Horizon, a "Study for the Establishment of a Lunar Military Outpost." The permanent outpost was predicted to cost $6 billion and was to become operational in December 1966 with twelve soldiers stationed at the Moon base. Does hacker Gary Mackinnon's discovery of defense department documents identifying "non-terrestrial officers" serving in space? Includes an 8-page color section.

346 Pages. 6x9 Paperback. Illustrated. $19.95. Code: HDAG

THE ANTI-GRAVITY FILES
A Compilation of Patents and Reports
Edited by David Hatcher Childress

With plenty of technical drawings and explanations, this book reveals suppressed technology that will change the world in ways we can only dream of. Chapters include: A Brief History of Anti-Gravity Patents; The Motionless Electromagnet Generator Patent; Mercury Anti-Gravity Gyros; The Tesla Pyramid Engine; Anti-Gravity Propulsion Dynamics; The Machines in Flight; More Anti-Gravity Patents; Death Rays Anyone?; The Unified Field Theory of Gravity; and tons more. Heavily illustrated. 4-page color section.

216 pages. 8x10 Paperback. Illustrated. $22.00. Code: AGF

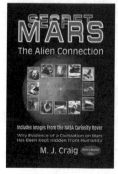

SECRET MARS: The Alien Connection
By M. J. Craig

While scientists spend billions of dollars confirming that microbes live in the Martian soil, people sitting at home on their computers studying the Mars images are making far more astounding discoveries... they have found the possible archaeological remains of an extraterrestrial civilization. Hard to believe? Well, this challenging book invites you to take a look at the astounding pictures yourself and make up your own mind. *Secret Mars* presents over 160 incredible images taken by American and European spacecraft that reveal possible evidence of a civilization that once lived, and may still live, on the planet Mars... powerful evidence that scientists are ignoring! A visual and fascinating book!

352 Pages. 6x9 Paperback. Illustrated. $19.95. Code: SMAR

ANCIENT ALIENS ON THE MOON
By Mike Bara
What did NASA find in their explorations of the solar system that they may have kept from the general public? How ancient really are these ruins on the Moon? Using official NASA and Russian photos of the Moon, Bara looks at vast cityscapes and domes in the Sinus Medii region as well as glass domes in the Crisium region. Bara also takes a detailed look at the mission of Apollo 17 and the case that this was a salvage mission, primarily concerned with investigating an opening into a massive hexagonal ruin near the landing site. Chapters include: The History of Lunar Anomalies; The Early 20th Century; Sinus Medii; To the Moon Alice!; Mare Crisium; Yes, Virginia, We Really Went to the Moon; Apollo 17; more. Tons of photos of the Moon examined for possible structures and other anomalies.
248 Pages. 6x9 Paperback. Illustrated.. $19.95. Code: AAOM

ANCIENT ALIENS ON MARS
By Mike Bara
Bara brings us this lavishly illustrated volume on alien structures on Mars. Was there once a vast, technologically advanced civilization on Mars, and did it leave evidence of its existence behind for humans to find eons later? Did these advanced extraterrestrial visitors vanish in a solar system wide cataclysm of their own making, only to make their way to Earth and start anew? Was Mars once as lush and green as the Earth, and teeming with life? Chapters include: War of the Worlds; The Mars Tidal Model; The Death of Mars; Cydonia and the Face on Mars; The Monuments of Mars; The Search for Life on Mars; The True Colors of Mars and The Pathfinder Sphinx; more. Color section.
252 Pages. 6x9 Paperback. Illustrated. $19.95. Code: AMAR

ANCIENT ALIENS ON MARS II
By Mike Bara
Using data acquired from sophisticated new scientific instruments like the Mars Odyssey THEMIS infrared imager, Bara shows that the region of Cydonia overlays a vast underground city full of enormous structures and devices that may still be operating. He peels back the layers of mystery to show images of tunnel systems, temples and ruins, and exposes the sophisticated NASA conspiracy designed to hide them. Bara also tackles the enigma of Mars' hollowed out moon Phobos, and exposes evidence that it is artificial. Long-held myths about Mars, including claims that it is protected by a sophisticated UFO defense system, are examined. Data from the Mars rovers Spirit, Opportunity and Curiosity are examined; everything from fossilized plants to mechanical debris is exposed in images taken directly from NASA's own archives.
294 Pages. 6x9 Paperback. Illustrated. $19.95. Code: AAM2

ANCIENT TECHNOLOGY IN PERU & BOLIVIA
By David Hatcher Childress
Childress speculates on the existence of a sunken city in Lake Titicaca and reveals new evidence that the Sumerians may have arrived in South America 4,000 years ago. He demonstrates that the use of "keystone cuts" with metal clamps poured into them to secure megalithic construction was an advanced technology used all over the world, from the Andes to Egypt, Greece and Southeast Asia. He maintains that only power tools could have made the intricate articulation and drill holes found in extremely hard granite and basalt blocks in Bolivia and Peru, and that the megalith builders had to have had advanced methods for moving and stacking gigantic blocks of stone, some weighing over 100 tons.
340 Pages. 6x9 Paperback. Illustrated.. $19.95 Code: ATP

THE TRIANGLE
The Truth Behind the World's Most Enduring Mystery
By Mike Bara

Ships have vanished without a trace only to magically reappear years later in good order but minus their crews, almost as if the intervening years had not even passed—for them. Compasses and guidance systems have spun inexplicably out of control over the shadowy waters of the Triangle. Entire squadrons of military aircraft have disappeared off of radarscopes in clear weather and with no forewarning. Explanations range from alien encounters to rogue waves to twisting unnatural funnel spouts caused by submerged civilizations left over from the days of Atlantis. Find out what really happened to Flight 19, the Navy training flight that last reported "they look like they're from outer space" over the Triangle.

218 Pages. 6x9 Paperback. Illustrated. $19.95. Code: TRI

ANCIENT ALIENS AND JFK
The Race to the Moon & the Kennedy Assassination
By Mike Bara

Relying on never-before-seen documents culled from the recent Kennedy assassination papers document dump, Bara shows the secret connections between key assassination figures like Oswald, LBJ, and highly placed figures inside NASA who had reasons to want Kennedy dead. Bara also looks into the bizarre billion-dollar Treasury bonds that Japanese businessmen attempted to deposit in a Swiss bank that had photos of Kennedy and the Moon on them. Is the wealth of the Moon being used as collateral by the USA? Bara digs into Kennedy's silent war with shadowy Deep State figures who were desperate to shut down his Disclosure agenda. Plus: "Apollo 20." Includes 8-page color section.

248 Pages. 6x9 Paperback. Illustrated. $19.95. Code: AAJK

THE ENCYCLOPEDIA OF MOON MYSTERIES
Secrets, Anomalies, Extraterrestrials and More
By Constance Victoria Briggs

Our moon is an enigma. The ancients viewed it as a light to guide them in the darkness, and a god to be worshipped. Did you know that: Aristotle and Plato wrote about a time when there was no Moon? Several of the NASA astronauts reported seeing UFOs while traveling to the Moon?; the Moon might be hollow?; Apollo 10 astronauts heard strange "space music" when traveling on the far side of the Moon?; strange and unexplained lights have been seen on the Moon for centuries?; there are said to be ruins of structures on the Moon?; there is an ancient tale that suggests that the first human was created on the Moon?; Tons more. Tons of illustrations with A to Z sections for easy reference and reading.

152 Pages. 7x10 Paperback. Illustrated. $19.95. Code: EOMM

OBELISKS: TOWERS OF POWER
The Mysterious Purpose of Obelisks
By David Hatcher Childress

Some obelisks weigh over 500 tons and are massive blocks of polished granite that would be extremely difficult to quarry and erect even with modern equipment. Why did ancient civilizations in Egypt, Ethiopia and elsewhere undertake the massive enterprise it would have been to erect a single obelisk, much less dozens of them? Were they energy towers that could receive or transmit energy? With discussions on Tesla's wireless power, and the use of obelisks as gigantic acupuncture needles for earth, Chapters include: Megaliths Around the World and their Purpose; The Crystal Towers of Egypt; The Obelisks of Ethiopia; Obelisks in Europe and Asia; Mysterious Obelisks in the Americas; The Terrible Crystal Towers of Atlantis; Tesla's Wireless Power Distribution System; Obelisks on the Moon; more. 8-page color section.

336 Pages. 6x9 Paperback. Illustrated. $22.00 Code: OBK

ORDER FORM

10% Discount When You Order 3 or More Items!

One Adventure Place
P.O. Box 74
Kempton, Illinois 60946
United States of America
Tel.: 815-253-6390 • Fax: 815-253-6300
Email: auphq@frontiernet.net
http://www.adventuresunlimitedpress.com

ORDERING INSTRUCTIONS

✓ Remit by USD$ Check, Money Order or Credit Card

✓ Visa, Master Card, Discover & AmEx Accepted

✓ Paypal Payments Can Be Made To:

   info@wexclub.com

✓ Prices May Change Without Notice

✓ 10% Discount for 3 or More Items

SHIPPING CHARGES

United States

✓ POSTAL BOOK RATE

✓ Postal Book Rate { $4.50 First Item
   50¢ Each Additional Item

✓ Priority Mail { $7.00 First Item
   $2.00 Each Additional Item

✓ UPS { $9.00 First Item (Minimum 5 Books)
   $1.50 Each Additional Item

   NOTE: UPS Delivery Available to Mainland USA Only

Canada

✓ Postal Air Mail { $19.00 First Item
   $3.00 Each Additional Item

✓ Personal Checks or Bank Drafts MUST BE

   US$ and Drawn on a US Bank

✓ Canadian Postal Money Orders OK

✓ Payment MUST BE US$

All Other Countries

✓ Sorry, No Surface Delivery!

✓ Postal Air Mail { $19.00 First Item
   $7.00 Each Additional Item

✓ Checks and Money Orders MUST BE US$
   and Drawn on a US Bank or branch.

✓ Paypal Payments Can Be Made in US$ To:
   info@wexclub.com

SPECIAL NOTES

✓ RETAILERS: Standard Discounts Available

✓ BACKORDERS: We Backorder all Out-of-
   Stock Items Unless Otherwise Requested

✓ PRO FORMA INVOICES: Available on Request

✓ DVD Return Policy: Replace defective DVDs only

ORDER ONLINE AT: www.adventuresunlimitedpress.com

10% Discount When You Order 3 or More Items!

Please check: ✓

☐ This is my first order   ☐ I have ordered before

Name

Address

City

State/Province | Postal Code

Country

Phone: Day | Evening

Fax | Email

Item Code	Item Description	Qty	Total

Please check: ✓

Subtotal ▶

Less Discount-10% for 3 or more items ▶

☐ Postal-Surface   Balance ▶

☐ Postal-Air Mail   Illinois Residents 6.25% Sales Tax ▶
   (Priority in USA)   Previous Credit ▶

☐ UPS   Shipping ▶
   (Mainland USA only)   Total (check/MO in USD$ only) ▶

☐ Visa/MasterCard/Discover/American Express

Card Number:

Expiration Date: | Security Code:

✓ SEND A CATALOG TO A FRIEND: